New Mathematical Pastimes

Written and Illustrated by

P A MacMahon

With a Contemporary Introduction by

Paul Garcia

Tarquin Books

Reprinting the 1930 Edition, Second Impression originally published by
Cambridge University Press

ISBN 1 899618 64 3

Tarquin Reprints distributed by QED Books, 195B Berkhamstead Road,
Chesham, HP5 3AP, United Kingdom

www.mathsite.co.uk

A catalogue record for this book is available from
the British Library.

Cover design by Jane Conway using an original illustration coloured by
John Sharp

Printed by Lightning Source in the UK and USA.

Contents

For a colour version of this book, there is a CD-ROM available. Contact details are set out at the back of this book.

Introduction

Paul Garcia

Introduction

Major Percy Alexander MacMahon is one of the nineteenth century's forgotten mathematical heroes (see Paul Garcia's biography on page vii). Mathematicians specialising in that area know him for his classic work Combinatory Analysis, a book still in print and relevant. He also published four papers on recreational topics: "On play á outrance" [1889], "Weighing by a series of weights" [1890], "On the thirty cubes constructed with six coloured squares" [1893], and "Magic squares and other problems upon a chess board" [1902].

MacMahon's less well-known book *New Mathematical Pastimes* develops some of the ideas in these recreational papers. It was published in 1921 with a second impression in 1930 soon after his death. It has had a widespread and persistent influence but is still unknown to many recreational mathematicians and the puzzle community. The book is hard to come by and so it seems fitting to produce a reprint to celebrate the 150th anniversary of his birth.

Historically, the book was important for a number of reasons. Unlike conventional recreational mathematics books, that were essentially collections of puzzles and their solutions, MacMahon set out to demonstrate the principles by which an interested reader, with some work and thought, could create his or her own amusements. It also contained a completely original approach to visual puzzles, where edge matching could be forced by altering the profiles of the edges, a process described in Part II of the book. The results of these transformations are remarkably similar to work produced by Maurits Escher, and predate him by a decade and a half. As well as being important as a resource the subject is by no means exhausted.

MacMahon obtained three patents for puzzles, in 1892 and 1893. The ideas in two of the patents, and the 1893 paper mentioned above, formed the core of *New Mathematical Pastimes*. One of these patents, for a cube puzzle based on the work in the 1893 paper, was turned into a commercial product, called Mayblox, sold in toy shops in the early twentieth century. Some puzzles based on the book were produced commercially in the 1930s and there are others still being produced. A description of these, together with a detailed description of MacMahon's other work in recreational mathematics by Paul Garcia can be found on page xi.

In the preface to *New Mathematical Pastimes*, MacMahon wrote "It has not been possible to produce the book in colour, and as the author has himself invariably investigated the sets in colour he must confess to a feeling of disappointment at the appearance of the pages." This lament is followed by a little quote:

> In colours fresh, originally bright,
> Preserve its portrait and report its fate!
> *The Complaint*

Unfortunately, this reprint is also in black and white, but there is a companion CDROM to this book which has a coloured electronic version of the book which brings it alive in a way we hope would have met MacMahon's approval. It also has some background material including pages to print to make your own puzzles.

I hope that a new generation might find inspiration and amusement in the puzzles that were an entirely original invention of this remarkable mathematician.

Paul Garcia
Long Melford
2004

Percy Alexander MacMahon (1854 – 1929)

MacMahon was a very unusual mathematician. He did not come from a family of mathematicians - his father and brothers were soldiers - and he did not have the university education normally associated with a prominent mathematician. Further, he came into the world of mathematics at a relatively late age (26), and did not really come to prominence until the age of 30. Until he joined the advanced class in mathematics at the Royal Military Academy in 1880 there had been no hint of any particular interest in, or talent for, doing mathematics, apart from an anecdote he told in later life about being interested in the stacking of cannon balls whilst a child in Malta.

Percy Alexander MacMahon was born in Sliema, Malta, on 26 September 1854, the second son of Brigadier-General Patrick William MacMahon and Ellen Curtis, daughter of George Savage Curtis of Teignmouth.

MacMahon attended the Proprietary School in Cheltenham. At the age of 14 he won a Junior Scholarship to Cheltenham College, which he attended as a day boy from 10 February 1868 until December 1870. MacMahon was always ranked in the top five in his classes, but never came first. At the age of 16 he was admitted to the Royal Military Academy at Woolwich, just a year after J. J. Sylvester had been forced to retire from the post of Professor. MacMahon later came to work in the same field as Sylvester, and there are some parallels between the careers of both men. He was eventually to write the obituary of Sylvester published in Nature in 1897 and in the Proceedings of the Royal Society in 1898 (extracts from which were used in the campaign to create the Royal Society's Sylvester medal).

After completing his training at Woolwich in 1873, MacMahon was posted to India as a Lieutenant in the Royal Artillery. As a result of illness, he was forced to return from India in 1878. Unable to return to active service, he enrolled in the Advanced Class for Artillery Officers. At the end of 1881 he was promoted to the rank of Captain and took up a post as an Instructor in Mathematics at the Royal Military Academy in Woolwich, which he held until 1889. During this period he married for first time and had a daughter, Florence.

Whilst employed as an instructor, MacMahon joined the London Mathematical Society (in 1883) and wrote a number of papers on a variety of algebraic topics, principally symmetric functions and invariants. In 1890, with the new rank of Major, he was appointed Professor of Physics at the Royal Artillery College, a post he held for eight years until his retirement from the army in 1898. During his tenure of this post he was elected a Fellow of the Royal Society (1890), the Royal Astronomical Society (1897), and held the Presidency of the London Mathematical Society from 1894 to 1896. He also received an honorary doctorate from Trinity College, Dublin. Although his academic mathematical output was considerable, in 1893 he wrote a paper entitled "On the thirty cubes that can be constructed with six coloured squares" which resulted in the book *New Mathematical Pastimes* in 1921. There were other papers on puzzles and geometrical designs (see page xix).

After the disappointment of his failure to be appointed Savilian Professor of Geometry at Oxford, MacMahon retired from the army in 1898 and threw himself into the mathematical and social life of London. He was involved with the British Association for Advancement of Science, as a member of the Council, President of Section A and then as General Secretary. The Royal Society awarded him the Royal Medal, Cambridge awarded him an honorary doctorate, and he joined the Athenaeum Club. He was also an active member of the Council of the London Mathematical Society. His mathematical work continued to be largely on the topic of partition theory, although he made excursions into other combinatorial areas such as magic squares.

In 1906 MacMahon took a post with the Board of Trade as Deputy Warden of the Standards, which brought him into contact with politicians such as David Lloyd George and Winston Churchill. In 1907, at the age of 53, he married Grace Elizabeth Howard, his second wife. Although his work at the Board of Trade was not particularly mathematical, and appears to have inspired no papers directly, over the next 15 years, MacMahon continued to produce papers on symmetric functions and partition theory, as well as writing four books (two of which, Combinatory Analysis I and II, first published in 1915 and 1916, are still in print today). Two further Universities, Aberdeen and St Andrews, awarded him honorary doctorates, he was President of the Royal Astronomical Society and the Royal Society awarded him the Sylvester Medal he had helped to create.

In 1922 MacMahon moved from London to Cambridge to continue the association he had had with St John's College Cambridge since 1904, when Professor Joseph Larmor had proposed on 18 November 1904 that MacMahon be invited to become a member of the College. He offered a lecture course with the title "Some processes in combinatory analysis", from 1923 until January 1925 and continued to lecture until 1928. MacMahon's mathematical interests expanded into more geometric fields, although the bulk of his still considerable output was on symmetric functions and related combinatorial problems. He received the De Morgan Medal of the London Mathematical Society in 1923, and in 1924 he attended and presented a paper at the International Mathematical Congress in Toronto.

In 1928 MacMahon's health had deteriorated and he moved to Bognor Regis on the advice of his doctors. He died there on Wednesday 25 December 1929, age 75.

MacMahon's contribution to mathematics

Although he never attended a University as a student, and only held an academic post at a traditional University at the end of his life, MacMahon was nevertheless a first class mathematician, popular and well-respected in his own lifetime, as evidenced by his three medals, four honorary degrees and involvement with a variety of prestigious societies, and whose legacy in the form of his books still in print and his Collected Works [Andrews 1985/86] is still cited by mathematicians today.

The speed with which MacMahon rose to prominence in the mathematical community of the late nineteenth century is remarkable, despite the fact that he was not a university man in the traditional sense. He was held in very high regard by many famous and well established mathematicians from the very start of his career; Sylvester, for example, specifically mentioned MacMahon three times in his inaugural speech as Savilian Professor of Geometry at Oxford. A review of MacMahon's two volume work, Combinatory Analysis [MacMahon, 1915/16], compared him with such luminaries as Fermat, Pascal, Euler, Lucas and Sylvester, at least in the field of combinatory analysis, and described his mathematical style as 'impeccable'.

MacMahon did pioneering work in invariant theory, symmetric function theory, and partition theory. Invariant theory deals with functions of the coefficients (and the variables) of expressions, that are invariant under a linear transformation of the variables. For example, in the quadratic expression, $ax^2+2bxy+cy^2$, the quantity $ac-b^2$ is invariant if x and y undergo a linear transformation (which simply shifts the origin of the graph of the expression to a different place). Symmetric functions are functions where interchanging any two of the variables leaves the function unchanged; for example, with four variables, $ab+ac+ad+bc+bd+cd$ is symmetric. Partition theory deals with counting the number of ways a number can be split into integers smaller than itself; for example, 3 can be split into 3, 2+1 and 1+1+1.

From this work he created the discipline we now call combinatorial analysis and he extended this to work in recreational mathematics, creating puzzles, but again in a pioneering, non-traditional way.

His summary works on Combinatorial Analysis [MacMahon 1915/1916], beginning with his 1896 Presidential address to the London Mathematical

Society, through his article in the tenth edition of the Encyclopaedia Britannica, to his three books on Combinatory Analysis are described in Chapter 8. The work on permutations in these books has resulted in certain statistics connected with the symmetric groups being called "Mahonian statistics." [Skandera, 2001] The partition theory developed in the books has resulted in there being more than 100 sequences in the "On-Line Encyclopedia of Integer Sequences" [Sloane] attributed in whole or part to MacMahon.

MacMahon also wrote on a variety of other combinatorial problems, including magic squares, repeating patterns, chess tournaments, voting systems (where his results were used in evidence to a Royal Commission in 1909), and weighing problems. These more 'recreational' aspects of his works are described on the following pages.

MacMahon's first major breakthrough was a paper in 1884 concerning semi-invariants and symmetric functions [Macmahon 1884] which brought him to the attention of the mathematical community, and only ten years later he was elected President of the London Mathematical Society. His work during that period was in partitions and invariant theory, and the link between them. In his Presidential address to the LMS in 1896, MacMahon drew attention the lack of any coherent work in the field of combinatorial analysis, which he regarded as the link between number theory (partitions) and algebra (invariants), the discontinuous and the continuous.

MacMahon was thus a pioneer and almost single-handedly created combinatorial analysis, by pursuing and combining two areas of work which had hitherto been regarded as separate. MacMahon was also a pioneer in applying this to recreational mathematics which is the subject of this book, again working in a field which was largely neglected by his contemporaries. He was not only an original and inventive mathematician, but also an historian of mathematics, an educator and populariser of mathematics, particularly in his involvement with the British Association for the Advancement of Science. He also provided encouragement for other mathematicians and supported the internationalisation of the mathematical community.

MacMahon's recreational mathematics

Although MacMahon's output on recreational topics is small compared to the volume of his writing on invariants and partitions it is important for several reasons. It shows the playful, non-serious side of MacMahon's character, and is one of his better known legacies from a general point of view. It also reveals MacMahon as a populariser of mathematics in *New Mathematical Pastimes*. In this work he is writing for a non-specialist audience and is trying to show how a mathematical way of thinking may be used to create interesting puzzles, designs and recreations. In the presentation of his work on repeating patterns to the Royal Society of Arts, he specifically set out to show to non-mathematicians how mathematical investigations might enhance the work of designers and artists.

Martin Gardner has written about MacMahon's work on three occasions ([Gardner 1966], [Gardner 1985], [Gardner 1992]), and 'MacMahon type' edge matching puzzles are still on sale today. Indeed, an internet search on "MacMahon" turns up several sites containing references to the MacMahon cube puzzle. I shall discuss some of the later references to MacMahon's recreational puzzles after I have described the papers working in chronological order. First, few preliminary remarks.

It is important to note that MacMahon was not a puzzle book writer in the style of Dudeney or Rouse-Ball. He was certainly aware of their work - it is listed in the bibliography to *New Mathematical Pastimes* - and both refer to his work, [Dudeney 1917 p109] [Rouse Ball] reference is made to MacMahon. The difference in approach was that MacMahon's purpose was not to provide entertainment, but to provide the means to make entertainment.

MacMahon also contributed a number of problems to the Educational Times, some of which are recreational in nature.

Early recreational papers

MacMahon's early papers on recreational mathematics do not fit with the main theme of MacMahon's recreational work, which was edge-matching puzzles and tilings (to use modern terminology).

"On play à outrance" [MacMahon 1889] considers the probabilities of the outcomes of a game for two players with a total of n counters between them.

"Weighing by a series of weights" [MacMahon 1890] introduces partitions in the context of producing sets of weights so that any integral weight from u lb. down to 1 lb. can be made in a unique way. A perfect partition (see note below) of u is needed to achieve this if weights may only be used on one side of a balance. If weights may be placed on both sides of a balance, then a sub-perfect partition will suffice; this is a partition in which the parts are allowed to be negative. This context presages his later work as Deputy Warden of the Standards for the Board of Trade (from 1906 to 1922).

Note: A perfect partition contains exactly one partition of every lower number (where number is taken to mean positive integer). For example, the partition (4, 2, 1) of 7 contains the partitions (4, 2) of 6, (4, 1) of 5, (4) of 4, (2, 1) of 3, (2) of 2 and (1) of 1.

MacMahon's patented puzzles

MacMahon was a creator of puzzles, some of which were manufactured and marketed, but it is unlikely that he made much money from his inventions.

The first patent for which MacMahon applied (number 21,118) was a puzzle comprising nine wooden blocks linked by flexible tapes to form a chain. The length of the links was to be such that the blocks could be formed into a stack in 4527 different ways. A design applied to the edges of the blocks would have to be reconstructed by finding the correct stacking. This patent was accepted on 8 October 1892. It is not known whether the puzzle was ever manufactured.

In 1892, MacMahon and his friend Major J. R. J. Jocelyn jointly applied for two patents. The first of these, patent number 3927, was for Appliances to be used in playing a new class of games. The provisional specification was submitted on 29 February 1892, and the patent granted on 28 January 1893. The patent is based on the equilateral triangle pastimes subsequently shown on page 2 of *New Mathematical Pastimes* using the pieces in figure 2. The rules of two different edge-matching games between two players to be played on specially marked boards were described in some detail, and a suggestion made for a puzzle where the pieces were to be fitted together to form a hexagon with (say) a single colour around the perimeter. It was this latter idea which was to be expanded upon in greater detail in *New Mathematical Pastimes* some 29 years later. It is not known whether any sets of pieces were manufactured or sold.

The second joint patent application was made on 2 May 1892. It described how a set of 27 coloured cubes could be constructed using three colours, the puzzle being to assemble the set into a larger cube with a uniform colour on each external face and matching between the internal faces. The existence of a set of 30 cubes made with six colours was also mentioned, and the puzzle described was to select two 'associated' cubes, which have the same colours on opposite faces, and then to locate amongst the remaining 28 cubes a set of sixteen in which none of those opposites occurs. These sixteen can then be assembled into two larger copies of the two associated cubes, where internal faces must also match.

It is this set of cubes that is described in the 1893 paper [MacMahon 1893], "On the thirty cubes constructed with coloured squares", which is the most important of the recreational papers, in the sense that it is the first public description of original recreational work by MacMahon. It was the second of two papers read at a meeting of the London Mathematical Society chaired by A. B. Kempe on 9 February 1893. The meeting and MacMahon's paper were reviewed in Nature on 23 February 1893.

In his paper, MacMahon first described how to calculate the number of different cubes it is possible to construct with six differently coloured squares, and then how to pair them up according to their symmetry group. From this basis, he described the essential puzzle: given any one of the thirty cubes (the target cube), to construct a cube of twice the linear dimensions with the same colour arrangement from eight of the remaining cubes, such that the internally touching faces match. This is done by using the octahedral dual of the selected cube (with the colours labelled 1 to 6) and performing a sequence of substitutions on the faces of the octahedron. The method selected the correct cubes and placed them in their correct orientations and positions. In *New Mathematical Pastimes*, the interested reader is referred to this paper for a detailed description of the method (see pages 42-47).

A puzzle called Mayblox was marketed by R. Journet and Co. of London (best known for their glass-topped dexterity puzzles) with the legend "Invented by Major P. A. MacMahon F. R. S." on the box. This version of the puzzle, described by Margaret A Farrell in the Journal of Recreational Mathematics in 1969 [Farrell, 1969], asked the solver to construct a large cube from 8 smaller cubes as in MacMahon's puzzle, but without the benefit of the target cube. Farrell described how to set about solving the puzzle, and also how to create a set of suitable cubes. A footnote to the article states:

> "The Mayblox puzzle was invented by Major P. A. MacMahon, R.R.S [sic] and was produced by a London firm, R. Journet. This puzzle was found among the effects of Maj. MacMahon's late aunt by Mr R. N. Andersen, a colleague at State University of New York at Albany."

This footnote is misleading, as explained by Professor Farrell in a personal communication to me, where she said that the footnote had been altered during editing to read "that the puzzle had been found in the effects of Major M[a]cMahon's aunt. It should have said, 'found in the effects of his late aunt...' meaning the effects of R. Andersen's late aunt." The particular puzzle in question was purchased in Hamley's of Regent Street, sometime between 1901 and 1915.

In 1930, Gerhard Kowalewski (Professor of Pure Mathematics at the Technische Hochschule in Dresden) wrote a chapter on the 'MacMahonsche' cubes in his Alte und Neue Mathematische Spiele [Kowalewski 1930]. The book gave instructions for constructing a set of cubes, with a detailed procedure for the colouring process. Kowalewski described the MacMahon version of the puzzle (and also claimed that a version of the puzzle was available commercially, in which eight cubes must be assembled to make a larger cube with faces each of a single colour - but without the benefit of the target cube, as in the Mayblox puzzle described above). Kowalewski also gave a version of his own where two sets of opposite faces have the same colour and the remaining two faces have a single colour each, so that two colours do not appear on the outside of the large cube.

In 1934, Ferdinand Winter, a student of Kowalewski's, published MacMahon's Problem - Das Spiel der 30 bunten Würfeln, a book reviewed, not entirely favourably, by W. L. Ayres [Ayres 1935] in 1935. The book discussed both MacMahon's puzzle and Kowalewski's variation. The level of detail was described by Ayres as tedious, which seems harsh given that the work was in fact Winter's thesis submitted to the Sachsische Technische Hochschule in Dresden for the degree of Doktor-Ingenieur. This is the list of contents (my translation):

1. Description of the puzzle
2. A new version of the puzzle
3. MacMahon's problem
4. The problem of two MacMahon cubes
5. Kowalewski's game with 16 cubes
6. The connection between MacMahon cubes and Kowalewski cubes
7. Locating the necessary cubes
8. A new game with three cubes
9. Cube puzzle: top and bottom the same colour
10. Cube puzzle: top and bottom with four colours
11. Cube puzzle with four colours on every outer face
12. Cubes with three hidden colours
13. Games with 12 and 16 cubes
14. Games with 16 cubes

15. A new game with 30 squares
16. A new game with magic cubes

Winter claimed that the puzzle was obtainable from the firm S. F. Fischer in Oberseiffenbach directly, or in toy shops and bookstores. There are no known surviving examples.

As part of Kowalewski's 60th birthday celebrations, the Monatshefte fuer Mathematik und Physik published a series of articles, including one by Winter [Winter, 1936] containing a new variant on the puzzle which involved building two 2 x 2 x 2 cubes such that only four colours appeared on the top and bottom faces, and the middle horizontal internal faces, with the vertical internal faces containing a repeating pair of the four colours, one pair in each cube. The remaining two colours had to appear in pairs on the side faces. Clearly Winter was pushing at the limits of the puzzle, turning it from an amusing pastime into an obsession. However, since MacMahon's purpose had been to involve the reader in actively creating puzzles by providing a mathematical framework, I think the project must be judged a success, certainly in Winter's case.

Magic squares

MacMahon's fourth recreational paper was "Magic Squares and other problems on a chessboard" [MacMahon 1902]. This was actually a transcript of a lecture given by MacMahon at the Royal Institution on 14 February 1902, published in its Proceedings and reprinted in Nature. MacMahon had been invited by the Royal Institution to present the talk at one of the Friday discourses for the general public. The session was chaired by Sir William de W. Abney, the Vice President of the R.I. at the time.

In his talk, MacMahon recounted some history of Magic squares and some methods of construction (in particular the method of De la Hire, where two Latin squares are added cell by cell), and then introduced the problem of enumerating the magic squares of a given order. This led seamlessly into a more general discussion of Latin squares, Graeco-Latin squares and group theory, and then to two famous problems: the "Probleme des Rencontres" ('in how may ways can n letters be placed in n envelopes so that no letter is in the correct envelope?'), as a way of calculating the number of ways the second row of a Latin Square might be completed given the first row, and thence to the "Probleme des Menages" ('in how many ways may n married couples be seated at a circular table so that no married pair sit together?') to calculate the number of possible third rows in the Latin Square.

The problem of placing eight castles on a chessboard so that none is attacked was used to provide a link between algebra (the study of continuous quantities) and arithmetic (the study of discrete quantities), by showing how the process of differentiation could be used to model the placing of the castles on successive rows of the chessboard.

The talk was illustrated with slides, and MacMahon was obviously aware of the non-specialist nature of his audience, for he gave clear explanations of terms and processes that a mathematical audience would not need. For example, in describing a variation of the Latin square where the number of symbols to be placed is greater than the number of cells in a row, so that there is no restriction on the number of symbols that may appear in each cell but each symbol must still appear once in each row and column, he claims that the number of arrangements is (4!)7. He gives this explanation of the notation: "4, the order of the square, must be multiplied by each lower number, and the number thus reached multiplied seven times by itself."

MacMahon's point was to show the audience that apparently widely disparate fields of study, such as the problem of the general nature of the magic square, and the infinitesimal calculus, could be linked to great advantage and reveal hidden depths to the processes involved.

New Mathematical Pastimes (NMP)

This book was first published in 1921, with a second impression printed in 1930, the year after MacMahon's death; in 1921 it cost 12 shillings. Albert A Bennett reviewed it for the *American Mathematical Monthly* in September 1922 [Bennett, 1922], describing it as "an interesting little volume filled with strange and bizarre figures, and punctuated with quaint quotations in verse." This is an accurate summary: since of the 114 pages, only 13 have no illustration, and there are 57 quotations scattered throughout.

The text is divided into three parts. The first part describes the triangle puzzles which were the subject of the patent number 3927 discussed above, and the extension of the idea to squares, right-angled triangles, pentagons and hexagons. Also described is the cube puzzle which was the subject of the 1893 paper, "On the thirty cubes constructed with coloured squares", discussed above. What makes the book different from other puzzle books is that it is not a collection of puzzles to be solved; it is instead a detailed explanation of how the puzzles may be constructed and solved, giving insight into how the art of mathematical thinking can be applied to a recreation.

An essential feature of the method is the subdivision of the shape in view (equilateral triangle, square, etc.) into (congruent) compartments, and then the matching of the compartments according to a rule, called a contact system. For example, the square can be divided into four compartments, numbered 1, 2, 3, 4, and then such squares can be arranged in a number of different ways: matching numbers (1 to 1, 2 to 2, etc.), or (1 to 2, 3 to 3, 4 to 4), or (1 to 2, 3 to 4) or (1 to 4, 2 to 3).

The second part of the book shows how the idea of edge-matching using colours can be replaced by altering the profile of adjoining edges to force the desired contact system. The method was used to produce some exotic jigsaw puzzles.

The third and final part of the book deals with the creation of repeating patterns. These would nowadays be described as tilings, beginning with the three regular tilings of the plane into equilateral triangles, squares or hexagons. From these bases, which MacMahon observed occur "constantly before our eyes", he describes how the methods developed in the first two parts of the book can be used to create an infinite variety of ways of subdividing the shapes and altering their boundaries according to certain rules. The role that symmetry has to play is emphasised, and there is a brief mention of the extension of the methods to three dimensions, where MacMahon explains how a space-filling array of cubes might be transformed into an array of rhombic dodecahedra. The importance of such matters to crystallography is also mentioned. In presenting this work to the world, MacMahon was ahead of his time.

Many of the planar designs in this part of the book are reminiscent of the designs of M. C. Escher (1898 - 1972), although they predate Escher's work by several decades. For example, pattern (b) on page 101 is described by Schattschneider in Tiling the plane with Congruent Pentagons [Schattschneider, 1978] as having "special aesthetic appeal". It is said to appear as street paving in Cairo; it is the cover illustration for Coxeter's Regular Complex Polytopes and was a favourite of the Dutch artist, M. C. Escher. Escher found the inspiration for it in a work of F. Haag from 1923 [Haag, 1923] in the Zeitschrift für Kristallographie, and it appears in his notebooks from the 1930s, described by Schattschneider in Visions of Symmetry [Schattschneider, 1990] on page 28.

The detail and symmetries of the construction of such patterns were taken up in greater depth in the papers described below.

The book concludes with a short set of recommendations concerning the construction of pieces for the puzzles, including suggested materials and colours.

Repeating patterns

Macmahon's suite of three papers on repeating patterns which expand upon themes from New Mathematical Pastimes were written or read in 1922, the year after New Mathematical Pastimes was published.

MacMahon's 1922 paper, "The design of repeating patterns - Part I", [MacMahon, 1922b] was coauthored with his nephew, William P. D. MacMahon, the son of his youngest brother, Reginald (1859 - 1911). W. P. D. MacMahon also published a paper in his own right, in 1925 [MacMahon, 1925], which is also described below. These two papers and the paper "The design of repeating patterns for decorative work" [MacMahon 1922a] which was read before the Royal Society of Arts on 10 May 1922, represent a unique body of work somewhat ahead of its time. According to Lockwood and MacMillan [Lockwood, 1978] although much work on space filling had been done by several people, from Kepler's work on sphere packing in 1611 [Kepler] to Fedorov and Schoenflies' enumeration of the space groups in 1890, little had been done on two-dimensional repeating patterns. Indeed, the concentration on three-dimensional crystallography meant that the 7 'frieze groups' and 17 'wallpaper groups' were not fully described until 1924 (by Polya). So MacMahon's interest in the subject at the time was unique for his time.

The purpose of the paper was "to establish a simple method for the design of repeats" and to introduce a calculus of symmetry. In the paper read for the Royal Society of Arts, this purpose is further expanded as "necessary to explore this field of thought as a preliminary to the study of the corresponding division of space of three dimensions which is required in the treatment of crystallography, crystal structure and the structure of the atom."

Indeed, Andrews [Andrews 1977/1986] observes that MacMahon's goal may been related to Hilbert's 18th problem, which was:

> Congress of Mathematicians at Paris in 1900: by Professor David Hilbert

> ### 18. Building up of Space from Congruent Polyhedra.

> If we enquire for those groups of motions in the plane for which a fundamental region exists, we obtain various answers, according as the plane considered is Riemann's (elliptic), Euclid's, or Lobachevsky's (hyperbolic) ...

> Now, while the results and methods of proof applicable to elliptic and hyperbolic space hold directly for n-dimensional space also, the generalisation of the theorem for Euclidean space seems to offer decided difficulties. The investigation of the following question is therefore desirable: Is there in n-dimensional Euclidean space also only a finite number of essentially different kinds of groups of motions with a fundamental region?

> A fundamental region of each group of motions, together with the congruent regions arising from the group, evidently fills up space completely. The question arises: Whether polyhedra also exist which do not appear as fundamental regions of groups of motions, by means of which nevertheless by a suitable juxtaposition of congruent copies a complete filling up of all space is possible. I point out the following question, related to the preceding one, and important to number theory and perhaps sometimes useful to physics and chemistry: How can one arrange most densely in space an infinite number of equal solids of given form, e.g., spheres with given radii or regular tetrahedra with given edges (or in prescribed position), that is, how can one so fit them together that the ratio of the filled to the unfilled space may be as great as possible?

There is evidence that other mathematicians also saw MacMahon's work in the context of Hilbert's problem. At a meeting of the Philadelphia Section of the Mathematical Association of America in February 1928, Professor A. H. Wilson of Haverford College presented a paper entitled "Space Filling Polyhedra", which began with an illustrated account of MacMahon's work on repeating polygons.

MacMahon's work on the subject stopped after the publication of the two papers I am about to describe, so that the 'Part II' implied by the title "The design of repeating patterns - Part I" was never written, for reasons described in detail in Andrews [Andrews 1986, Chapter 15]. MacMahon wrote to Baker on 5 February 1922 describing a space-filling tetrahedron he had discovered. On the letter, held in St. John's College, there is a pencilled note, presumably by Baker, "?Is this one of Schoenflies Crystallosysteme p299." Andrews felt that this had dampened MacMahon's enthusiasm for the subject. Baker was correct. On page 299 of Schoenflies' *Krystallsysteme und Krystallstructur* [Schoenflies, 1891] there is a description of a construction which produces the same tetrahedron as MacMahon's. Schoenflies had started with a square based prism with a length twice as long as the side of the square base. He then joined the midpoints of three adjacent faces to one another and to the vertex of the prism at the intersection of the faces. The result was a tetrahedron with four identical isosceles faces, where the ratio of the long side to the short side was 1:½√3 should go in here. MacMahon did not indicate how he had discovered it. It is pity that MacMahon was apparently dissuaded from further investigation of this object, since it has been 'rediscovered' by puzzle enthusiasts and used to create, for example, rotating rings of joined tetrahedra, which I am sure would have delighted MacMahon. In his letter to Baker, MacMahon wrote: "It is most fruitful in further results. I have made a rough model and am having some made by a pattern maker." This suggests that MacMahon was intending to pursue the possibilities afforded by this polyhedron, and may have done so without getting around to publishing his results.

In "The design of repeating patterns - Part I", MacMahon presented a series of definitions and results which provided a language for talking about repeating patterns, since, as was made clear in the RSA paper, extensive investigations into both artistic and mathematical literature had revealed a lack of development in the field. MacMahon thus clearly felt, quite justifiably, that he was pioneering a new area of work.

Modern treatments of the subject of tiling are essentially descriptive, whereas MacMahon's approach was constructive. By using the subdivision of a polygon and the rules of contact (edge matching) as developed from the original patent of 1892 in NMP, MacMahon was able to show how tilings of the plane with specific symmetries could be created from triangles, squares and hexagons. For example, the pattern from page 101 of NMP is derived from a square with contact system 1 to 4, 2 to 3.

MacMahon was also prepared to consider designs in which the repeating unit contained holes ('stencil' repeats) or comprised disconnected pieces 'archipelago' repeats). The nomenclature is ascribed to his Cambridge friend, the geometer G. T. Bennett so it seems clear that MacMahon felt that the topic was important enough to discuss with eminent colleagues.

The calculus of symmetry MacMahon developed was based on the contact systems between the polygons, and the rules of symmetry which must apply to the edges in contact in order to force the desired contact system. MacMahon identified two types of symmetry for the edge contacts: mirror symmetry in a line at right angles to the untransformed edge, and point symmetry about the centre point of the untransformed edge, illustrated below (the diagrams are adapted from MacMahon's own [MacMahon 1922a]).

Untransformed contact
edge along AB

Contact with point
symmetry about centre of
contact edge

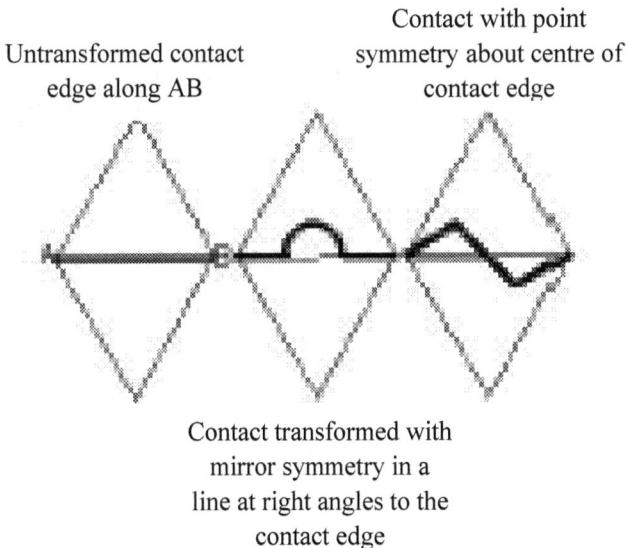

Contact transformed with
mirror symmetry in a
line at right angles to the
contact edge

The symbol I was used to indicate mirror symmetry, P to indicate point symmetry.

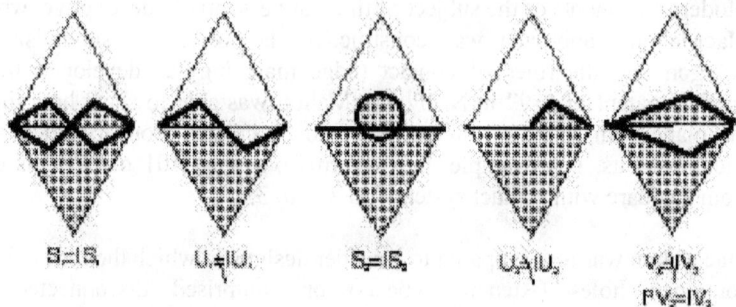

S=IS₁   U₁=IU₂   S₂=IS₂   U₂=IU₁   V₂=IV₂
                                      PV₂=IV₁

The suffices 1 and 2 were used to indicate the nature of the contact: 1 for same colours touching, 2 otherwise. The diagram above illustrates the five classes of edge transformations identified in the paper. The notations S, U and V denote transformations which are symmetrical about the line at right angles to the original contact edge, those which have point symmetry about the midpoint of original contact edge and those which have neither, respectively. The equations under the diagrams show that, for example, an S transformation is the same as its mirror image IS; or for a V transformation, the point symmetry image is the same as the mirror symmetry image (PV = IV).

Assemblages of the shapes according to the contact systems create repeat units which are simply translated on the plane to fill it. MacMahon identified the possible symmetries of repeat units, but did not seek to enumerate all the possible symmetries derivable from those symmetries. It may be that this was an idea he intended to pursue in the unwritten part II.

MacMahon does not seem to have considered in detail the enumeration of contact systems available to him. For example, in the case of the square divided into four compartments numbered 1, 2, 3, 4 in a clockwise order illustrated above, there are ten different contact systems available (the ones MacMahon had illustrated in *New Mathematical Pastimes* are indicated with the page reference and figure number):

{11,22,33,44} (NMP page 92 fig 110b),
{12,33,44},
{13, 22,44},
{14,22,33},
{11, 23,44},

{11,24,33}, (NMP page 96 fig 116)
{11,22,34},
{12,34},
{13,24}, (NMP page 99 fig 120a)
{14,23}.

Of these, only those printed in bold are achievable without changing the order of the numbers within the square. MacMahon did not provide examples of all the possible contact systems. The contacts refer to the colour numbers, so {11,24,33} means colour 1 is always adjacent to colour 1, colour 2 adjacent to colour 4 and 3 adjacent to 3 – in this example you have to extend the tiling across the plane to see the 11 contacts.

With 5 numbered compartments in a pentagon, there are 26 potential contact systems, and 6 compartments in a hexagon provides 76 possibilities. The sequence generated by the enumeration of possible contact systems is the same as the number of self-inverse permutations on n letters and may be found by searching the On-line Encyclopaedia of integer sequences [Sloane] for the sequence 1, 2, 4, 10, 26, 76,...

The direction in which MacMahon was going is clear: from the consideration of edge-matching the coloured tiles, through the replacement of the colours by transformed contact edges to the creation of repeat units, he has created a fundamental region in the Hilbert sense, which will generate the tiling from its symmetries.

The paper stopped rather abruptly, and the baton was passed to the nephew, W.P. D. MacMahon. In his paper [MacMahon, 1925], W. P. D. MacMahon explained how he had been involved with his uncle in "an investigation of the design of repeating patterns in two and three dimensions, a subject the idea of which was first broached by Major MacMahon in a little book entitled New Mathematical Pastimes, recently published." He went on to say, "the author believes that many of the results in the present paper are new...", evidence that both MacMahons believed they were breaking new ground.

The paper started with a definition of a polygon or 'repeat', set out the labelling conventions for the diagrams and defined the idea of a 'system of contact' that formed the backbone of *New Mathematical Pastimes*. These polygons could be either convex or concave (that is, one or more of the angles could be re-entrant) - for the quadrilateral this was a new result. From these fundamental ideas, the 'Law of Angle Distribution' was derived: a closed polygon may be a repeat if, and only if, its angles are such that they may be associated together in sets of two, three, or four differently lettered angles, the sum of the angles in each set being equal to π or 2π.

From this, W. P. D. MacMahon was able to produce a table showing how the associations referred to in the theorem must be distributed. All triangles and quadrilaterals are repeats, but for a pentagon to be a repeat, it must have a set of three angles that sum to 2p and the remaining two must sum to p, or vice versa. The possible kinds of contact system possible for repeat pentagons and hexagons were considered in some detail.

The next, and last, paper in this series was the paper read to the Royal Society of Arts on 10 May 1922 [MacMahon 1922a]. This must have been a very entertaining talk, illustrated with many slides and pictures 'thrown upon' a screen, probably using an epidiascope. The main body of the talk described in outline the content of New Mathematical Pastimes and the papers referred to above, but what was remarkable about the talk was its audience, which included many artists. MacMahon's opening remarks provided a justification for the study of two-dimensional repeating patterns as a precursor to the study of the three dimensional case, which was of great importance to crystallography and the study of the structure of the atom. He described how he had conducted a search of geometrical and art literature before commencing his investigation, and then stated his purpose in making a presentation to the Royal Society of Arts "to bring to the notice of those who are concerned with the practical use of patterns the results that have been arrived at for scientific purposes."

The main claim was that the method of edge transformation and the calculus of symmetry he had developed was sufficient to create and classify all possible repeating patterns. In the report of the discussion afterward, it was clear that some members of the audience were not entirely convinced of this, and also cast some doubt on the aesthetic qualities of some of the patterns shown by MacMahon.

Apart from a letter in Nature in 1922 [MacMahon, 1922c] and a note in the Proceedings of the London Mathematical Society in 1924 [MacMahon, 1924], mentioned but not reproduced in Andrews, there is no evidence that MacMahon continued to pursue this line of work, which had occupied him on and off for 30 years, possibly for the reason already mentioned.

Later developments

Many people have taken the puzzles invented by MacMahon further. The paper by Farrell on Mayblox has already been described above. The articles by Martin Gardner mentioned above first appeared in *Scientific American* and were subsequently included in the books to which reference has been made. In the discussion which follows, I have ignored references to, and descriptions of,

'Instant Insanity' type puzzles, in which a column or tower of cubes must be constructed according to certain rules, since there is no evidence that MacMahon was ever concerned with such puzzles.

The first article [Gardner, 1966] described the 24 three-colour squares, and the cubes. In it, Gardner told of how the squares problem led to an article in New Scientist in 1961 [O'Beirne, 1961], an enumeration of 12 224 different patterns possible with the squares by Federico Fink in Buenos Aires and an exhaustive computer search in 1964 (which found 12 261 patterns).

The second article [Gardner, 1985], reproduced in Mathematical Magic Show, dealt with the 24 four-colour triangles, and mentioned the work of Wade Philpott (1918 - 1985), an engineer from Ohio. Philpott became interested in MacMahon's puzzles during a period of hospitalisation following a shooting accident in 1947. His archive is now stored in the University of Calgary (as part of the Eugene Strens Recreational Mathematics Collection). Wade Philpott created versions of the square and triangle puzzles that he called 'Multimatch.' These are made and sold today by Kate Jones of Kadon Enterprises; there is correspondence between Philpott and Kate Jones from the period 1981 to 1984 in the Calgary archive. A set was exhibited at the conference on recreational mathematics in 1986 to celebrate the founding of the Strens Collection. In the book of the conference [Guy 1994] this is described as a set originally produced by Philpott in cardboard accompanied by a construction problem "that Percy never thought of". Kadon now produce a jumbo size in laser cut acrylic.

The final article [Gardner, 1992] dealt exclusively with the 30 coloured cubes. This is the only one of the three articles to include a bibliography. In 1956, Paul B. Johnson [Johnson, 1956] wrote an article about the cube puzzle in which he discussed the number of ways of selecting 8 cubes to form a duplicate of a randomly selected 'key cube' from the full set of thirty possible cubes, and the number of ways in which possible sets of cubes will 'stack' to form the required duplicate. He ignored the requirement that internal faces must also match. This article also has a bibliography.

In 1971, Norman T. Gridgeman [Gridgeman, 1971], in The 23 colored cubes, took a wider view of the question of colouring cubes by allowing the cubes to be coloured in a single colour, two colours, three colours, etc., up to six different colours. These he grouped into 23 species (so in that sense the title may be misleading - it would more accurately the 23 species of coloured cubes – but only to avoid confusion with MacMahon's 30 coloured cubes). This article also includes a bibliography.

Two articles in the early 1970s [Kahan 1972] [Sobczyk 1974] describe a variant of the Mayblox puzzle, invented by an Irishman, Eric Cross, which had recently been put on the market under the name "Eight Block to Madness". The eight blocks had to be assembled to form a larger cube after the manner of the Mayblox, but without the requirement that internal faces match.

References

[Andrews 1977/1986]
George Andrews, *Collected Works of Percy Alexander MacMahon*, published in two volumes in 1977 and 1986
Where the papers by MacMahon are referenced in Andrews, they are shown by a note such as [Andrews: 33:10] which means it is the 33rd paper by MacMahon and may be found in Chapter 10.

[Ayres 1935]
W. L. Ayres, Review of [Winter 1930] *American Mathematical Monthly* 1935: 42:563-564

[Bennett, 1922],
A. A. Bennett, Reviews: New Mathematical Pastimes, *American Mathematical Monthly* September 1922, 29:307-309 1922

[Dudeney 1917]
Amusements in Mathematics London 1917, republished 1970 as a Dover edition

[Farrell, 1969]
Margaret A Farrell, "The Mayblox problem" in the *Journal of Recreational Mathematics* 1969 pp51 - 56

[Fedorov 1891a]
E S Federov: "Symmetry of Regular Systems of Figures". *Proceedings of the Imperial Saint Petersburg Society Series 2*, 28:1-146
[Fedorov 1891b]
E S Federov 1891b: "Symmetry in the Plane." *Proceedings of the Imperial Saint Petersburg Society Series 2*, 28: 345-89

[Gardner 1966]
New Mathematical Diversions, Unwin 1966

[Gardner 1985]
Mathematical Magic Show

[Gardner 1992]
Fractal Music, Hypercards and more

[Gridgeman, 1971]
Norman T. Gridgeman "The 23 colored cubes", *Mathematics Magazine*, 44(5), November 1971, pp243 - 252

[Guy 1994]
Richard K. Guy and Robert E. Woodrow, editors, *The Lighter Side of Mathematics*, Mathematical Association of America 1994

[Haag, 1923]
F. Haag, "Die regelmässingen Planteilungen und Punktesystem", *Zeitschrift für Kristallographie*, 1923 58:478-88

[Johnson 1956]
Paul B. Johnson, "Stacking coloured cubes", *American Mathematical Monthly*, 63, June 1956, pp392 - 395

[Kahan 1972]
Steven J. Kahan , "Eight blocks to madness - a logical solution", *Mathematics Magazine*, Vol. 45 1972 pp57 – 65

[Kowalewski 1930]
Gerhard Kowalewski, *Alte und Neue Mathematische Spiele* 1930

[Lockwood, 1978]
E. H. Lockwood and R. H. MacMillan, *Geometric Symmetry*, CUP 1978

[MacMahon 1886]
P. A. MacMahon, "Certain special partitions of numbers" *Quarterly Journal of Mathematics* 1886: XXI: 367 – 373 (Andrews [20:6])

[MacMahon 1889]
P. A. MacMahon, "On play "à outrance"", *Proceedings of the London Mathematical Society* 1889: XX. 195-198 [Andrews: 33:10]

[MacMahon 1890]
P. A. MacMahon, "Weighing by a series of weights" *Nature* 1890: XLII . 113-114 [Andrews: 37:19]

[MacMahon 1893]
P. A. MacMahon, "On the thirty cubes constructed with six coloured squares" Proceedings of the London Mathematical Society, 1893, 24, pp145 – 155 [Andrews: 44:15]

[MacMahon 1902]
P. A. MacMahon, "Magic squares and other problems upon a chess board"

Proceedings of the Royal Institution (Volume XVII 96, February 1902, pp 50 - 61) reprinted in Nature in March 1902 [Andrews: 60:19]

[MacMahon 1915/1916]
P. A. MacMahon:, *Combinatory analysis Volume 1*, 1915, *Volume 2* 1916, currently available as a Chelsea reprint

[MacMahon 1922a]
P. A. MacMahon: "The design of repeating patterns for decorative work." *Journal of the Royal Society of Arts* 70, 1922, 567-78 and related discussion 578-82 [Andrews:99:15]

[MacMahon, 1922b]
P. A. MacMahon and W. P. D. MacMahon: "The design of repeating patterns Part 1". *Proceedings of the Royal Society of London* 101, 1922, 80-94 [Andrews: 98:15]

[MacMahon, 1922c]
P. A. MacMahon, "Pythagoras's theorem as a repeating pattern" *Nature*, 109, 1922, p479 [Andrews: 100:15]

[MacMahon, 1925]
W. P. D. MacMahon: "The theory of closed repeating polygons in Euclidean space of two dimensions" *Proceedings of the London Mathematical Society* (2) 23 1925, 75-93

[Rouse-Ball]
W.W. Rouse Ball, *Mathematical Recreations and Essays*, this was originally published in 1892 and went through numerous editions including the twelfth edited by H. S. M. Coxter in 1974. MacMahon is mentioned in all editions. A 1942 reprint of the 11th edition contains a number of references to MacMahon. The first is in connection with the 'Probleme des Menages' (p 50) and directs the reader to *Combinatory Analysis Vol 1* (1915). The second (p51) is to MacMahon's paper, "Weighing by a series of weights." The third reference in Rouse-Ball (p111) is to the edge matching puzzles in *New Mathematical Pastimes*, using an illustration from NMP, but in the 12th this is replaced by a discussion on pentominoes and MacMahon is no longer mentioned. Coxeter also loses the references to *Combinatory Analysis* concerning the problem of placing Queens on a chessboard.

[Schattschneider 1978]
D. Schattschneider in "Tiling the plane with Congruent Pentagons", *Mathematics Magazine*, 51, 1978, 29-44

[Schattschneider, 1990]
Doris Schattschneider, *Visions of Symmetry, Notebooks Periodic drawings and related work of M. C. Escher*, Freeman New York 1990

[Schoenflies 1891]
Schoenflies, *Kristallsysteme und Kristallstructur Teuber* Leipzig 1891

[Skandera, 2001]
"An Eulerian partner for inversions", *Seminaire Lotharingien de Combintaoire*, 46, 2001 for a clear outline of these statistics.

[Sloane]
The On-Line Encyclopedia of Integer Sequences can be found on the web at: www.research.att.com/~njas/sequences/ It was originally published as tables: N. J. A. Sloane, *A handbook of Integer Sequences*, Academic Press 1973. The book does not contain any references to MacMahon.

[Sobczyk 1974]
Andrew Sobczyk, "More progress to madness via eight blocks", *Mathematics Magazine*, Vol. 47, 1974, pp115 - 124

[Winter 1930]
Ferdinand Winter, MacMahon's Problem - Das Spiel der 30 bunten Würfeln 1930 published by Teubner. There are three known copies of this work: one at The University of California in Berkeley, one at the University of Cincinnati, and one in the Niedersachsische Staats und Univeritaets library at Goettingen.

[Winter, 1936]
Ferdinand Winter, "Eine neue Aufgabe zu MacMahons 30 bunten" *Wuerfeln Monatshefte fuer Mathematik und Physik* 44,1936 (Wirtinger), pp290 -294

NEW
MATHEMATICAL
PASTIMES

BY

Major P. A. MACMAHON, R.A.

D.Sc., Sc.D., LL.D., F.R.S.
ST JOHN'S COLLEGE, CAMBRIDGE

CAMBRIDGE
AT THE UNIVERSITY PRESS
1921

... ; he put together a piece of joinery, so crossly indented and whimsically dove-tailed; a cabinet so variously inlaid; such a piece of diversified mosaic; such a tessellated pavement without cement; ...

EDMUND BURKE, 1774, on *American Taxation.*

PREFACE

EDWARDS. You are a philosopher, Dr Johnson. I have tried too in my time to be a philosopher, but I don't know how, cheerfulness was always breaking in.

BOSWELL'S *Life of Johnson.*

THE author of this book has, of recent years, devoted much time and thought to the development of the subject of 'Permutations and Combinations' with which all students are familiar. He has been led, during that time, to construct, for use in the home circle, various sets of pieces, of elementary geometrical shapes based upon these ideas, and he now for the first time brings them together with the object of introducing, in a wider sphere, what he believes to be a pleasant by-path of mathematics which has almost entirely escaped the attention of the well-known writers upon Mathematical Recreations and Amusements. The book differs *in toto* from their works because everything that it contains, with scarcely an exception, is the invention of the author.

It is not a bringing together of materials derived from wholly different ideas. From beginning to end it proceeds along one defined path from which it never diverges. One continuous thread of thought runs through it from cover to cover.

Indeed, in view of the excellent collections of mathematical recreations that have proceeded from the pens of Ed. Lucas, W. W. Rouse Ball, W. Ahrens, H. E. Dudeney and others it would appear that there is no room for another book upon the lines upon which these have been written. I would make particular reference to the work in two volumes of W. Ahrens of Magdeburg and draw attention to the Bibliography which it contains. It involves nearly eight hundred titles. This must have necessitated much research. He has included a large number of works upon magic squares and upon the knight's tour and other chess-board amusements. These have been omitted from the short list appended to this book, which had been almost completed when the list of Ahrens came into view. Many of the books appear to be rare, as they could not be found in the library of the British Museum.

Part I deals with sets of pieces of exactly the same size and shape, but differently coloured, numbered or otherwise distinguished upon combinatory principles in such wise that no two pieces of a set are identical. It is shewn that such sets lend themselves to a great variety of pastimes, and the reader who will take the trouble to construct sets and employ suitable colours will find that he is truly in a kaleidoscope of constantly changing colour effect, the attractiveness of which will fully repay him for his trouble. The designs obtained from some of the contact systems frequently possess beautiful symmetry.

It has not been found possible to produce the book in colour, and as the author has himself invariably investigated the sets in colours he must confess to a feeling of disappointment at the appearance of the pages.

> In colours fresh, originally bright,
> Preserve its portrait and report its fate!
>
> *The Complaint.*

Briefly, Part I may be described as generalised dominoes.

Part II follows Part I in a natural manner, for it is a mere transformation of it. The sets of Part I have the same shape but are differently coloured. Those of Part II have the same colours but are differently shaped. The transformations can be carried out always in an infinite number of ways and give ample scope for taste and ingenuity. One merely requires squared paper (in the millimetre unit by preference), ruler and compasses to be able to design and construct a means of endless amusement.

> How rich the prospect! and for ever new!
> And newest to the man who views it most;
> For newer still in infinite succeeds.
>
> *The Consolation.*

Part III follows quite naturally upon Part II because the schemes of transformation prove, after a little consideration, to be a comprehensive method of designing repeating patterns for decorative work. Here we have pieces of the same size and shape which can be employed to completely cover a pavement or other flat surface. The only repeating patterns that were known in early times appear to have been the equilateral triangle, square and regular hexagon, the only regular polygons which possess the 'repeating' property. Great advances were made in the middle ages by the Arabian and Moorish architects, and in many of their buildings—*ex. gr.* the Alhambra at Granada— elaborate repeating patterns, based upon the square and its derivatives, are in evidence and are most effective. Each of these can be labelled with base and contact system in the classification in this book. Repeating patterns are in constant view in the home in parquet floors, carpets, paper-hangings, apparel, woven fabrics, etc.

An attempt has been made to interest the reader while drawing as little as may be upon his knowledge of geometry. If he will study Part III after having become familiar with the transformations of Part II he will be able to make designs for home work to his heart's content.

Beauty of form as depending upon symmetry of some kind is brought forward as being necessarily an important object of the designer. An elementary discussion of this precedes definite rules for obtaining symmetry.

The fact that the number of different patterns is unlimited, in each of the categories, leaves much to the judgment and fancy of the designer who can give free play to his imagination.

The patterns usually met with in public edifices and private homes are on hard and fast lines, shewing curiously little variety and cleverness in view of the fact that the theme is one of infinite scope.

The subject may be regarded, it is thought, as an important recreation because the construction of the designs and assemblages possesses a distinct fascination.

> What we admire we praise; and when we praise
> Advance it into notice, that its worth
> Acknowledg'd, others may admire it too.
> *The Task.*

The subject of Part III has been carried much further than appears in these recreations, by the author and others. A work, entirely devoted to it, is in hand and may shortly appear.

P. A. M.

September, 1921.

TABLE OF CONTENTS

PART I

PASTIMES BASED UPON SIMPLE GEOMETRICAL FORMS

PART II

THE TRANSFORMATION OF PART I

PART III

THE DESIGN OF 'REPEATING PATTERNS' FOR DECORATIVE WORK

PART I

PASTIMES BASED UPON SIMPLE GEOMETRICAL FORMS.

> Come track with me this little vagrant rill,
> Wandering its wild course from the mountain's breast.
>
> DOUBLEDAY.

1. The amusements to which this book is devoted are played with a number of cards or pieces which involve or are based upon certain regular or other polygons which are distinguished upon the sides with certain colours or numbers, in such wise that for a given number of colours and for given conditions of their occurrence there is one piece for every possible arrangement of the colours upon the sides.

The principal existing amusement which embodies this idea

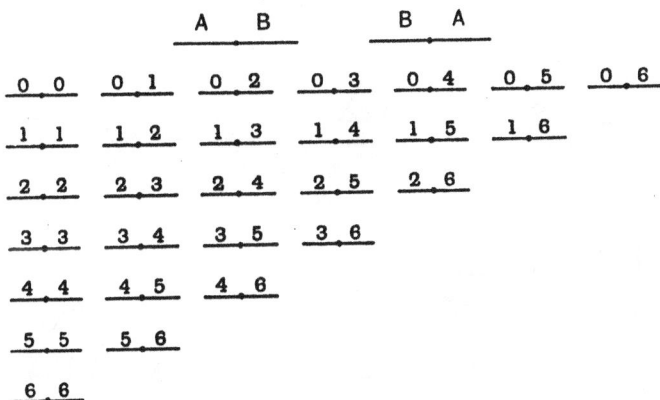

Fig. I.

is the game of dominoes which in various varieties is known all over the world.

The pieces here consist essentially of straight lines—each piece being a line as AB in fig. I.

In the most common set seven numbers are employed

$$0, 1, 2, 3, 4, 5, 6$$

and these are arranged in every possible way upon the two

halves *A, B* in such wise that the pieces *A—B* and *B—A* (fig. 1) are regarded as identical.

We thus obtain 28 pieces as numbered in fig. 1.

For convenience, in actual practice, the pieces are broadened so as to consist of two equal squares joined about a side, each square being devoted to a number.

Dudeney, Lucas and others have shewn that, apart from the different games, Draw, Matadore, Cyprus, etc., the pieces may be placed together so as to fulfil certain conditions and thus lend themselves to a great variety of patiences and puzzles.

Other sets of dominoes have been popular at different times and in different places; such for instance as employ ten or thirteen different numbers so as to proceed to the double-nine and double-twelve pieces respectively.

EQUILATERAL TRIANGLE PASTIMES

A scheme, analogy pronounced so true:
Analogy, man's surest guide below.

The Complaint.

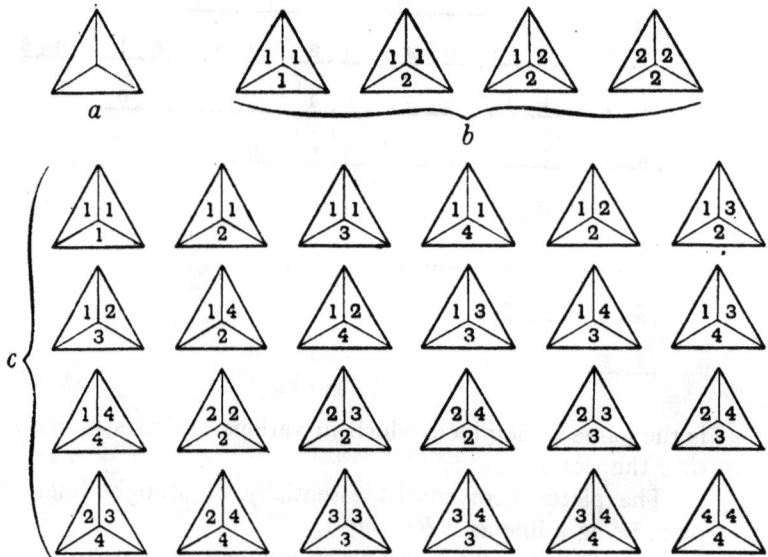

Fig. 2.

2. In order to develop this principle take an equilateral triangle (fig. 2*a*) and join the angular points to the centre of the circumscribing circle. This point is also the point of intersection of the three perpendiculars from the angular points upon the opposite sides and each perpendicular has the point as one of its points of trisection. The construction divides the triangle into three equal and similar parts. These parts we will call compartments of the triangle.

If we employ two colours, denoting them by the numbers 1, 2 and, as in dominoes, allow repetitions of the same numbers on the same triangle we obtain the four pieces of fig. 2*b*.

Three colours yield eleven and four colours twenty-four pieces*. The twenty-four pieces are set forth arranged in a convenient order in fig. 2*c*.

---

* The problem of enumerating the number of ways of colouring the sides of a regular polygon of m sides with n colours, repetitions of colour on the sides of the polygon being permitted, may be studied by means of the Theory of Cyclical Permutations which has been very clearly set forth by Netto in his *Combinatorik* (Leipzig, Verlag von B. G. Teubner, 1901). If upon a set of polygons the n colours are repeated α, β, γ, ... times respectively, where any of the numbers α, β, γ, ... may be zero and α, β, γ, ... do not contain any common factor greater than unity, the number of different polygons of the set is

$$\frac{(n-1)!}{\alpha!\,\beta!\,\gamma!\,...}, \quad (n!\text{ means the factorial of } n, \text{ sometimes written } \underline{n})$$

but if the numbers α, β, γ, ... involve a common factor d, where d is a prime number, and $\alpha = d\alpha_1$, $\beta = d\beta_1$, $\gamma = d\gamma_1$, ... $n = dn_1$, the number of different polygons of the set is

$$\frac{(n-1)!}{\alpha!\,\beta!\,\gamma!\,...} + \frac{n-n_1}{n} \cdot \frac{(n_1-1)!}{\alpha_1!\,\beta_1!\,\gamma_1!\,...}.$$

The results are more complicated when the common divisor d is a composite number but are given (l.c.).

Utilising these results we find that the enumerations are for

Triangle	$\frac{1}{3}n\,(n^2 + 2),$
Square	$\frac{1}{4}n\,(n+1)\,(n^2 - n + 2),$
Pentagon	$\frac{1}{5}n\,(n^4 + 4),$
Hexagon	$\frac{1}{6}n\,(n+1)\,(n^4 - n^3 + n^2 + 2),$
Heptagon	$\frac{1}{7}n\,(n^6 + 6).$

The mathematical reader will be able to establish that when the polygon has p sides, p being a prime number, the enumeration is given by

$$\frac{1}{p}n\,(n^{p-1} + p - 1).$$

When repetitions of colour are not permitted the enumeration, in the case of a regular polygon of m sides, is

$$\frac{1}{m}\frac{n!}{(n-m)!}.$$

If one particular colour is to occur upon each polygon the enumeration is given by

$$\frac{(n-1)!}{(n-m)!}.$$

These may be set up into a regular hexagon as in fig. 3, a circumstance which supplies a useful starting-point for the study of the system of triangles.

When we assemble the pieces so as to form the hexagon we may adopt some principle of contact between the compartments of different triangles. The most obvious one is to insist that a

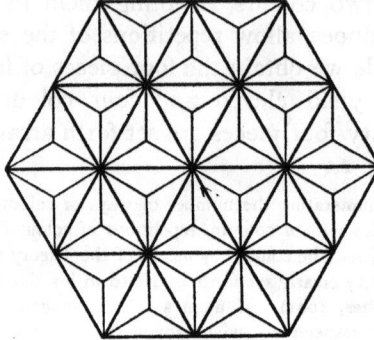

Fig. 3.

compartment shall lie adjacent to a compartment similarly numbered or coloured. With this single condition the number of arrangements is so large that it proves to be possible to impose other conditions affecting the compartments which lie upon the exterior boundary of the hexagon.

Meditation here
May think down hours to moments.
The Task.

In order to realise what may be the nature of these boundary conditions we must observe that since each of the 24 triangles possesses 3 compartments there are altogether 72 compartments. The four colours are symmetrically involved so that each must appear upon one-fourth of 72 or upon 18 compartments. By reason of the specified contact condition a particular colour must occur an even number of times inside the hexagon and since 18 is an even number it follows that a particular colour must occur an even number of times upon the boundary. We see that there are 12 boundary compartments so that any particular colour must occur upon the boundary a number of times denoted by one of the even numbers

0, 2, 4, 6, 8, 10, 12.

Thence it follows that, considering all four colours, the occurrence upon the boundary must be according to one of the eight schemes:

Colours	1	2	3	4		1	2	3	4
	12	0	0	0		6	4	2	0
	10	2	0	0		6	2	2	2
	8	4	0	0		4	4	4	0
	8	2	2	0		4	4	2	2

The fifth of these types indicates that the colours occur upon the boundary 6, 4, 2, 0 times respectively and the type is looked upon as the same when the colours are merely interchanged; so

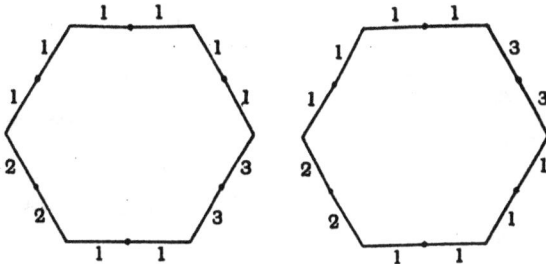

Fig. 4.

that the 24 arrangements of the numbers 6, 4, 2, 0 are all of the same type.

Each type except the first, in which all the boundary compartments are of the same colour, involves several varieties of

C1,1,1,1 B12,0,0,0   C1,1,1,1 B6,6,0,0

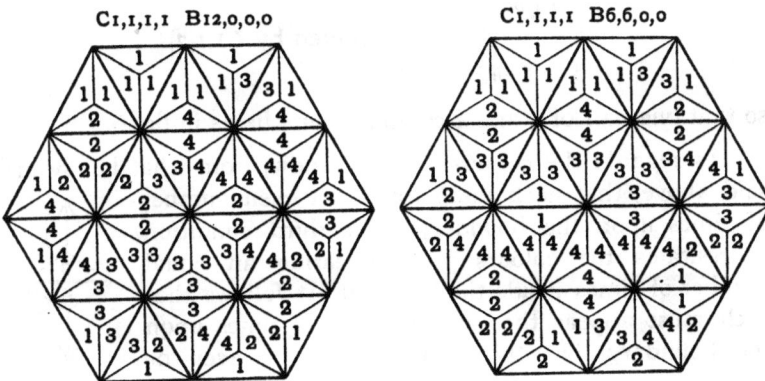

Fig. 5.

boundary depending upon the way in which the given boundary colours are arranged.

Thus for the fourth type two varieties are as in fig. 4.

With the stated condition of contact inside the hexagon each of the eight types can be actually set up and many varieties of types. Some of these are difficult to obtain and make considerable demands upon the skill and patience. It is not known if any of the varieties are in reality impossible boundary conditions.

Two examples are given in fig. 5.

They are framed upon what we will call the First Contact System, which we denote by $C_{1,1,1,1}$.

> So from the first eternal order ran,
> And creature link'd to creature, man to man....
> ....
> The link dissolves, each seeks a fresh embrace,
> Another love succeeds, another race.
>
> · *Essay on Man.*

3. We may adopt other contact conditions inside the hexagon. We have four colours at disposal, 1, 2, 3, 4 suppose. In the first contact system we have

$$\left.\begin{array}{ccc} 1 \text{ adjacent to } & 1 \\ 2 & \text{\textit{,,}} & 2 \\ 3 & \text{\textit{,,}} & 3 \\ 4 & \text{\textit{,,}} & 4 \end{array}\right\} \text{ denoted by } C_{1,1,1,1}.$$

We take as our second contact system

$$\left.\begin{array}{ccc} 1 \text{ adjacent to } & 1 \\ 2 & \text{\textit{,,}} & 2 \\ 3 & \text{\textit{,,}} & 4 \end{array}\right\} \text{ denoted by } C_{1,1,2},$$

so that one pair of 3 and 4 compartments lie as in fig. 6.

As regards the colours 1, 2, we know, from what has been said above, that each must appear an even number of times upon the boundary. The third condition necessitates the colours 3, 4 appearing an equal number of times inside the hexagon. It follows that each must appear the same number of times upon the boundary, but this number may be even or uneven.

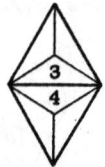

Fig. 6.

The possible types of boundary are now 16 in number:

Colours 1	2	3	4	1	2	3	4
12	O	O	O	6	2	2	2
10	2	O	O	4	4	2	2
8	4	O	O	6	O	3	3
6	6	O	O	4	2	3	3
10	O	1	1	4	O	4	4
8	2	1	1	2	2	4	4
6	4	1	1	2	O	5	5
8	O	2	2	O	O	6	6

and there are varieties of every type except the first.

Most of these types and many varieties have been actually set up. Probably every type is possible but nothing is known

C1,1,2 B6,6,0,0   C1,1,2 B12,0,0,0

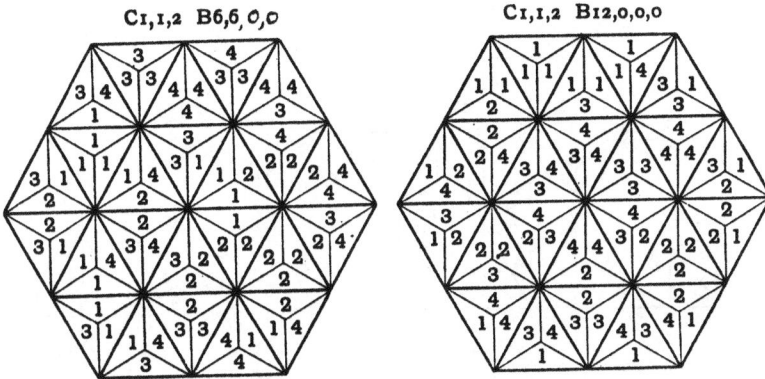

Fig. 7.

concerning the possibility of the numerous varieties except that it is certain that a great number of these can be arranged.

Two examples are given in fig. 7.

> Arithmetick would erre exceedingly,
> Forgetting to divide and multiply;
> Geometry would lose the altitude,
> The crassie longitude and latitude;
> And musick in poore case would be o'er throwne,
> But that the goose quill pricks the lessons downe.
>   TAYLOR'S *Works*, 1630.

4. For a third contact system we take

1 adjacent to 2

3   „   4

that is to say the colours 1, 3 lie up against the colours 2, 4 respectively. We denote it by $C2,2$.

From what has been said above it is clear that the colours 1, 2 must occur the same number of times upon the boundary and also the colours 3, 4.

The types are four in number:

Colours	1	2	3	4
	6	6	0	0
	5	5	1	1
	4	4	2	2
	3	3	3	3

and every type possesses varieties.

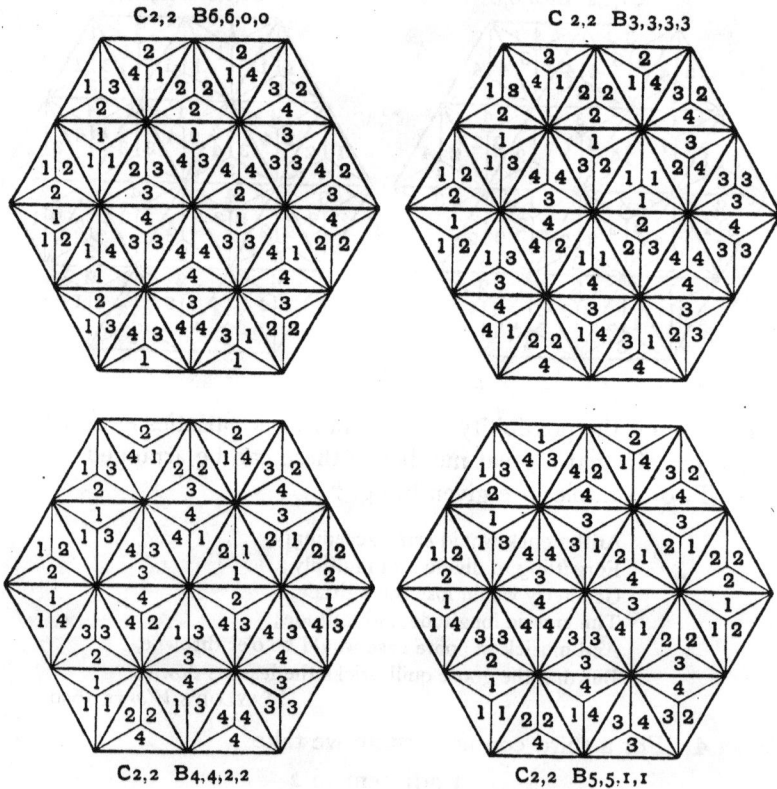

Fig. 8.

It will be noticed that since the triangles are symmetrical in the four colours, 2, 2 | 4, 4 is not regarded as a type distinct from 4, 4 | 2, 2.

These types and many varieties can be set up.

As to the possibility of particular varieties the usual doubt remains.

A number of examples are given in fig. 8.

5. We have thus, in the case of four colours, three systems of conditions of inside contact.

It is convenient to denote these by

$$C_{1,1,1,1}, \quad C_{1,1,2}, \quad C_{2,2}.$$

For any number of colours a system is determined by the number of pairs of colours that are associated. The number of systems is the number of ways in which the number, which expresses how many colours are in use, can be made up of ones and twos.

For   1, 2, 3, 4, 5, 6, 7, 8, 9, 10, ... colours

the systems are   1, 2, 2, 3, 3, 4, 4, 5, 5, 6, ... in number.

In other words, to obtain the number of systems we add 1 or 2 to the number of colours, according as that number is uneven or even, and then divide by 2.

The scheme of triangles that has just been studied may be realised also by taking instead of four colours the blank and one, two, three pips upon the compartments. It then constitutes a

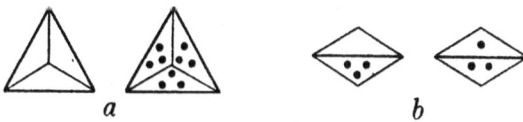

a   *b*

Fig. 9.

set of triangular dominoes proceeding from treble blank to treble three as in fig. 9*a*.

The first contact system is similar to that adopted for the ordinary game of dominoes, whilst the third system is of the nature of that obtaining in the game of Matadore.

In fact for the third system we may take, as in fig. 9*b*,

blank compartment adjacent to three-pip,

one-pip   „   „   two-pip,

so that the condition has the statement that the sum of the pips in adjacent compartments must be three.

Good mother, how shall we find a pig if we do not look about for it? Will it run off o' the spit, into our mouths, think you as in Lubberland, and cry we we?   B. JONSON, *Barth. Fair*, III. 2.

6. We can select, from the complete set of 24, less numerous sets which are interesting to examine.

There is a set of 11 triangles which involves only three specified colours as in fig. 10, but this set cannot be assembled into any interesting shape; so that we attempt to frame an additional condition with the object of reducing the number of triangles to 10.

It will be noticed that 10 out of the set of 11 are altered by the interchange of the colours 1 and 2. The triangle which is

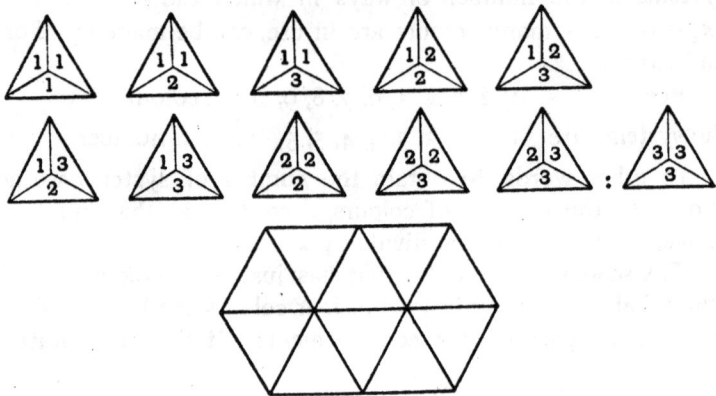

Fig. 10.

not altered by the interchange is the one which has all three compartments with the colour 3.

We therefore make a new condition to the effect that every triangle is to be altered by the interchange of the colours 1, 2.

We thus obtain 10 pieces which can be put together into the hexagonal shape in the lower part of fig. 10.

It is important to bear in mind that this set is not symmetrical in three colours, but notwithstanding that there is only symmetry in regard to the colours 1, 2 the set is interesting, as will appear.

Examination of the 10 pieces shews that colours 1, 2, 3 occur upon 11, 11, 8 compartments respectively.

There are eight boundary compartments.

For the contact system $C1,1,1$, since each of the colours must occur an even number of times inside, it follows that upon the boundary the colours 1, 2, 3 must occur numbers of times which are uneven, uneven, even respectively.

We obtain the types, six in number:

Colours	1	2	3		1	2	3
	7	1	0		3	3	2
	5	3	0		3	1	4
	5	1	2		1	1	6

and varieties of each type.

Examples are given in fig. 11.

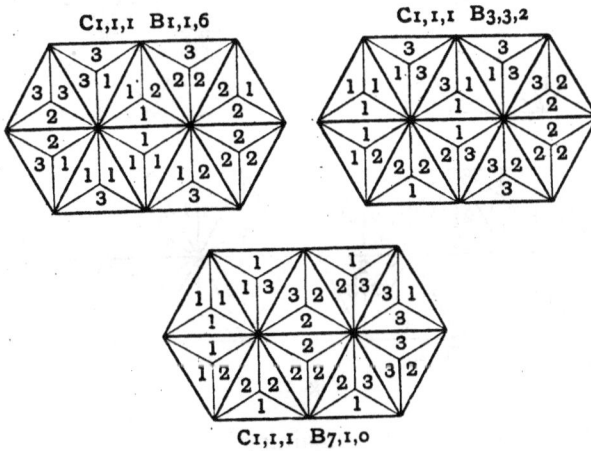

C1,1,1 B1,1,6   C1,1,1 B3,3,2

C1,1,1 B7,1,0

Fig. 11.

The contact system in which 1 is adjacent to 2 and 3 adjacent to 3 is different from that in which 1 is adjacent to 1 and 2 to 3 because of the want of symmetry in these colours. We will denote these by $C2,1$ and $C1,2$ respectively.

7. In the case of $C2,1$ the colours 1, 2 must occur an equal number of times on the boundary and the colour 3 an even number of times not exceeding 6. The limit 6 is necessary because only 6 of the 10 pieces involve the colour 3.

We have now four types:

Colours	1	2	3
	4	4	0
	3	3	2
	2	2	4
	1	1	6

and varieties as usual.

Examples are given in fig. 12.

An example of *every* type is given as an indication that probably in other pastimes all types are possible. The reader however will notice that other considerations may rule out certain types, as in fact is seen to be the case in this particular pastime.

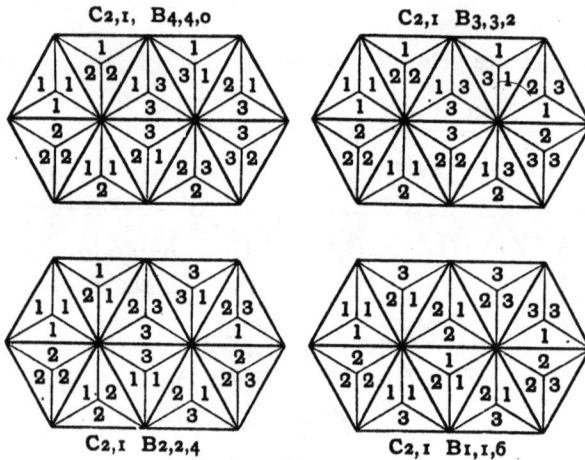

Fig. 12.

8. In the case of the remaining contact system $C1,2$, viz. 1 to 1 and 2 to 3, we find colour 1 must occur an uneven number of times upon the boundary while colour 2, since it occurs inside just as often as colour 3, must occur three more times on the boundary than colour 3.

Three types arise:

Colours	1	2	3
	1	5	2
	3	4	1
	5	3	0

Examples are given in fig. 13.

Here we see that more than one variety of the type $B5,3,0$ can be arranged.

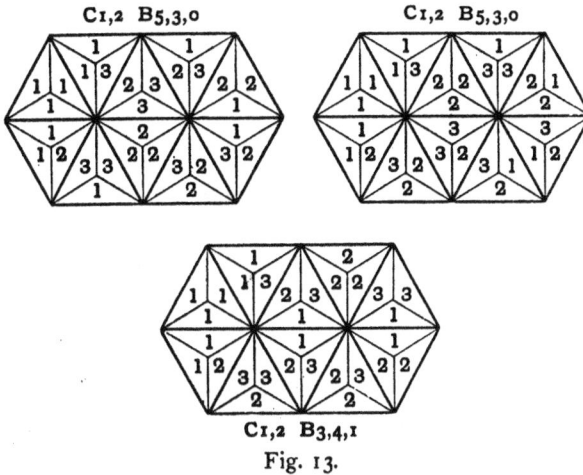

Fig. 13.

9. From the complete set of 24 triangles we now isolate the set containing all of those which involve one particular colour, say the colour 4. There are thirteen as in fig. 14 which can be

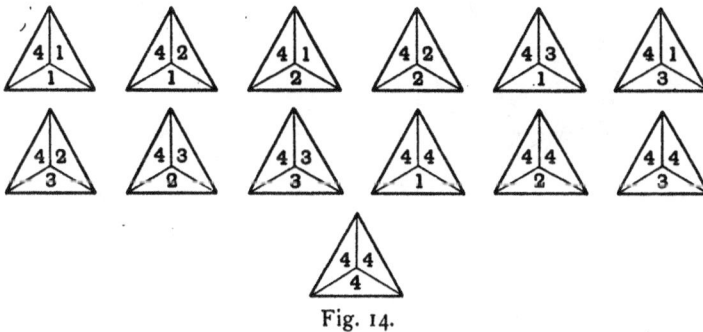

Fig. 14.

assembled into the form of a semi-regular hexagon or blunted triangle as in fig. 15.

Observe that the set is symmetrical in the *three* colours 1, 2, 3 but that it is not symmetrical in four colours.

There are nine boundary compartments.

The colours 1, 2, 3, 4 occur in 7, 7, 7, 18 compartments respectively.

For the contact system $C1,1,1,1$ the boundary conditions

are that the colours 1, 2, 3 must each occur an uneven number of times and 4 an even number.

There are thus six types:

Colours 1	2	3	4		1	2	3	4
1	1	1	6		5	1	1	2
3	1	1	4		5	3	1	0
3	3	1	2		3	3	3	0

and the usual varieties.

The type 7, 1, 1, 0 is seen at a glance to be impossible. Examples are given in fig. 15.

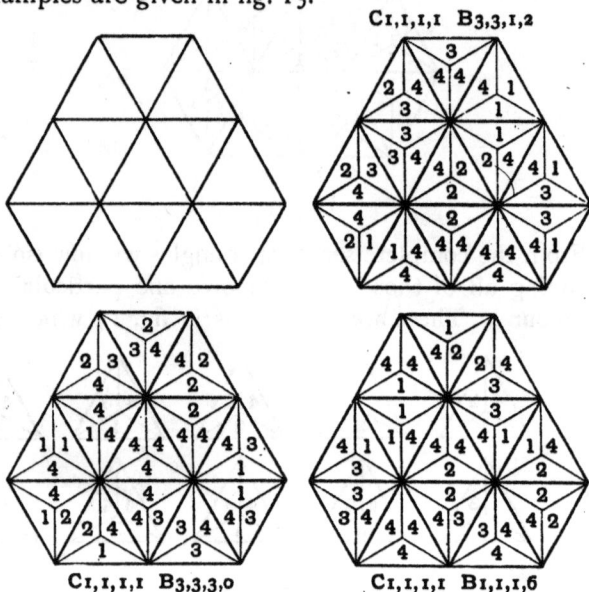

Fig. 15.

To guide my course aright
What mound or steady mere is offered to my sight?

DRAYTON, *Polyolb.* 1. p. 569.

10. For the contact system $C1,1,2$, viz.:

$$1 \text{ to } 1,$$
$$2 \text{ ,, } 2,$$
$$3 \text{ ,, } 4,$$

we find as one boundary condition that the colour 4 must occur on the boundary 11 more times than colour 3, an impossibility since there are only nine boundary compartments. Hence the system $C1,1,2$ is impossible.

For the system $C1,2,1$, viz. :

$$1 \text{ to } 1,$$
$$2 \text{ ,, } 3,$$
$$4 \text{ ,, } 4,$$

we find as the boundary conditions

1 must occur an uneven number of times,
2 as many times as 3,
4 an even number of times,

leading to the types, ten in number :

Colours 1	2	3	4		1	2	3	4
1	1	1	6		3	2	2	2
3	0	0	6		5	1	1	2
1	2	2	4		1	4	4	0
3	1	1	4		3	3	3	0
1	3	3	2		5	2	2	0

Examples are given in fig. 16.

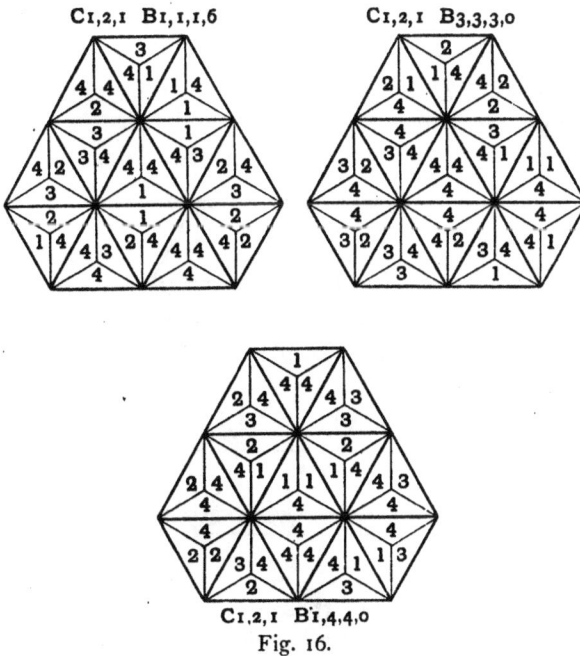

$C1,2,1$ $B1,1,6$ $C1,2,1$ $B3,3,3,0$

$C1,2,1$ $B1,4,4,0$
Fig. 16.

This band dismiss'd, behold another crowd
Preferr'd the same request, and lowly bow'd.
                                    The Temple of Fame.

11. The next set of triangles is concerned with the equilateral triangle and five colours, three different colours appearing
in the compartments of each triangle. This design leads to the

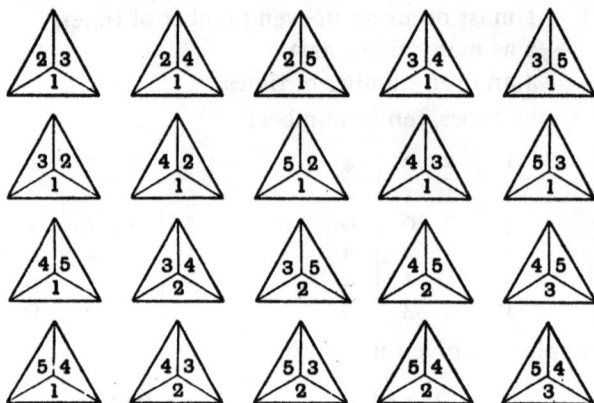

Fig. 17.

twenty triangles, as in fig. 17, which can be arranged in the
semi-regular decagon of fig. 18.

There are 12 boundary compartments. Altogether there are

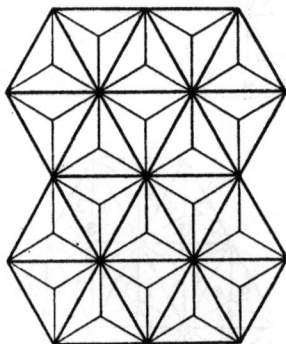

Fig. 18.

60 compartments and 12 of these must belong to each colour as
each of the five colours is symmetrically involved.

For the contact system C1,1,1,1,1 which is

$$
\begin{aligned}
&1 \text{ to } 1, \\
&2 \text{ „ } 2, \\
&3 \text{ „ } 3, \\
&4 \text{ „ } 4, \\
&5 \text{ „ } 5,
\end{aligned}
$$

we gather that each colour must occur an even number of times upon the boundary.

We have therefore the types, ten in number:

Colours	1	2	3	4	5	1	2	3	4	5
	12	0	0	0	0	6	4	2	0	0
	10	2	0	0	0	6	2	2	2	0
	8	4	0	0	0	4	4	4	0	0
	8	2	2	0	0	4	4	2	2	0
	6	6	0	0	0	4	2	2	2	2

with varieties of each type except the first.

The existence of the first type is somewhat remarkable as there are only just sufficient compartments of a particular colour to go completely round the boundary.

It will be noted that the 20 triangles are naturally arranged

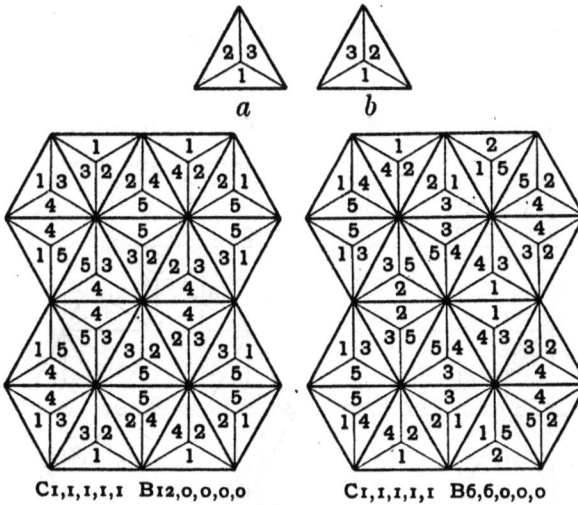

C1,1,1,1,1 B12,0,0,0,0 C1,1,1,1,1 B6,6,0,0,0

Fig. 19.

in pairs, because to every triangle (as fig. 19a) there can be associated the complementary (as fig. 19b).

The latter, read counterclockwise, is the same as the former read clockwise. This circumstance supplies a clue to the setting up of some of the boundary types. It is left to the reader to take advantage of this hint.

Some examples are given in fig. 19.

12. For the contact system C1,1,1,2, viz. 1 to 1, 2 to 2, 3 to 3, 4 to 5, the colours 1, 2, 3 must each occur an even number of times upon the boundary and the colours 4, 5 the same number of times.

The boundary types are 23 in number:

Colours 1	2	3	4	5	1	2	3	4	5	1	2	3	4	5
12	0	0	0	0	8	2	0	1	1	6	0	0	3	3
10	2	0	0	0	6	4	0	1	1	4	2	0	3	3
8	4	0	0	0	6	2	2	1	1	2	2	2	3	3
8	2	2	0	0	4	4	2	1	1	4	0	0	4	4
6	6	0	0	0	8	0	0	2	2	2	2	0	4	4
6	4	2	0	0	6	2	0	2	2	2	0	0	5	5
4	4	4	0	0	4	4	0	2	2	0	0	0	6	6
10	0	0	1	1	4	2	2	2	2					

with varieties of every type except the first.

Examples are given in fig. 20.

C1,1,1,2 B0,0,0,6,6   C1,1,1,2 B12,0,0,0,0

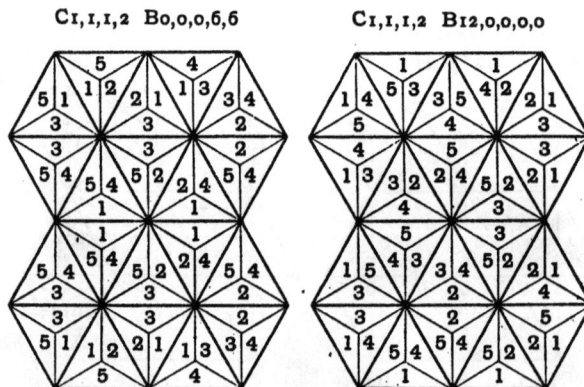

Fig. 20.

13. The remaining contact system $C_{1,2,2}$, viz. 1 to 1, 2 to 3, 4 to 5, yields the sixteen types:

Colours 1	2	3	4	5		1	2	3	4	5		1	2	3	4	5
12	0	0	0	0		4	4	4	0	0		2	3	3	2	2
10	1	1	0	0		4	3	3	1	1		0	6	6	0	0
8	2	2	0	0		4	2	2	2	2		0	5	5	1	1
8	1	1	1	1		2	5	5	0	0		0	4	4	2	2
6	3	3	0	0		2	4	4	1	1		0	3	3	3	3
6	2	2	1	1												

with varieties of each type except the first.

Examples, which include two varieties of $B_{0,3,3,3,3}$, are given in fig. 21.

$C_{1,2,2}$ $B_{12,0,0,0,0}$

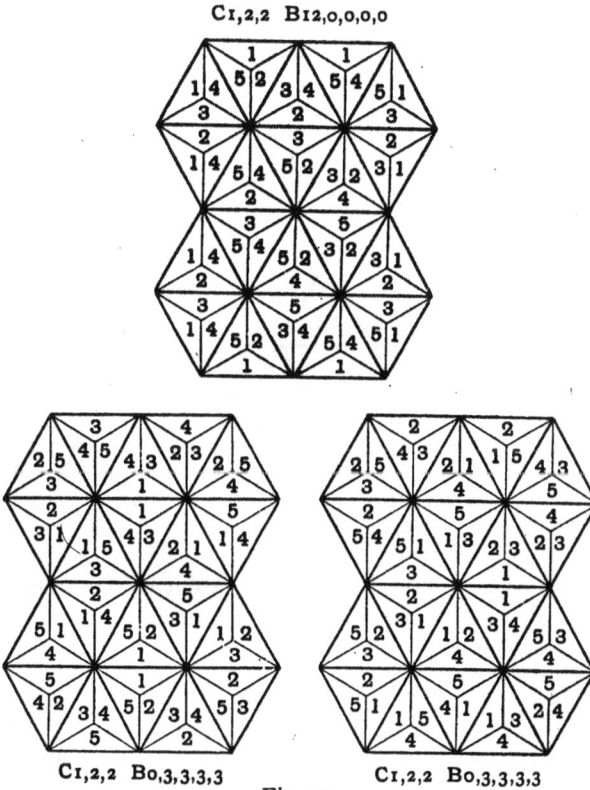

$C_{1,2,2}$ $B_{0,3,3,3}$ $C_{1,2,2}$ $B_{0,3,3,3,3}$

Fig. 21.

This set is prolific of types, the three systems of contact yielding altogether 49.

Experiments with these will be found by many to be very interesting.

Eum odi sapientem qui sibi non sapit:
- hee is an ill cooke that cannot licke his own fingers.

WITHALS' *Dict.*, Ed. 1634.

14. We can divide the complete set into two portions:

(i) those which involve a particular colour, twelve in number;

(ii) the remainder, eight in number, which involve only four colours 1, 2, 3, 4.

The first set of 12 is symmetrical in the four colours 1, 2, 3, 4.

There are 36 compartments which occur 6, 6, 6, 6, 12 times for the five colours 1, 2, 3, 4, 5 respectively.

The members of the set may be assembled in either of the forms of fig. 22, each of which has 10 boundary compartments.

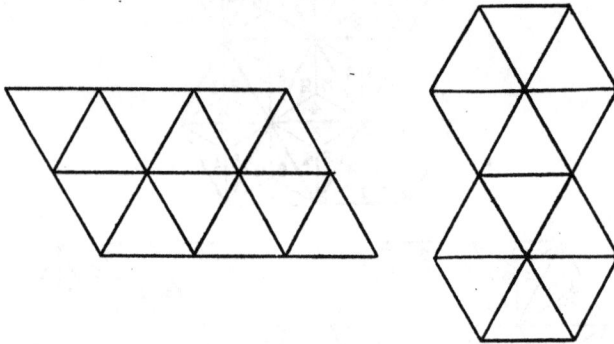

Fig. 22.

For the contact system $C1,1,1,1,1$ each colour must appear an even number of times upon the boundary, so that the types are fifteen in number:

Colours	1	2	3	4	5		1	2	3	4	5		1	2	3	4	5
	0	0	0	0	10		4	2	0	0	4		2	2	2	2	2
	2	0	0	0	8		2	2	2	0	4		6	4	0	0	0
	4	0	0	0	6		6	2	0	0	2		6	2	2	0	0
	2	2	0	0	6		4	4	0	0	2		4	4	2	0	0
	6	0	0	0	4		4	2	2	0	2		4	2	2	2	0

Examples are given in fig. 23.

Certain of these types exist for only one of the two boundary shapes, as it is seen at once that the first of the types cannot exist for the parallelogram, while it does exist for the double hexagon.

$C1,1,1,1$ $B10,0,0,0,0$ $C1,1,1,1$ $B4,4,2,0,0$ $C2,1,1,1$ $B0,0,2,0,8$

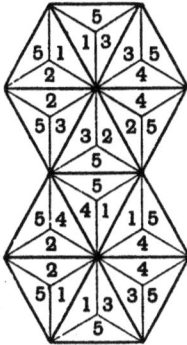 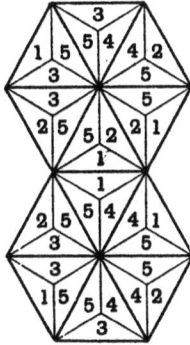 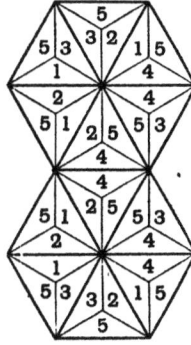

Fig. 23.   Fig. 24.

15. For the contact system $C2,1,1,1$ we find that the colours 1, 2 must occur equally often and the colours 3, 4, 5 each an even number of times. The types are 30 in number:

Colours	1	2	3	4	5	1	2	3	4	5	1	2	3	4	5
	5	5	0	0	0	2	2	4	0	2	1	1	4	4	0
	4	4	2	0	0	2	2	2	2	2	0	0	0	0	10
	4	4	0	0	2	2	2	2	0	4	0	0	2	0	8
	3	3	4	0	0	1	1	0	0	8	0	0	4	0	6
	3	3	0	0	4	1	1	2	0	6	0	0	2	2	6
	3	3	2	2	0	1	1	4	0	4	0	0	6	0	4
	3	3	0	2	2	1	1	2	2	4	0	0	4	2	4
	2	2	6	0	0	1	1	6	0	2	0	0	6	2	2
	2	2	0	0	6	1	1	4	2	2	0	0	4	4	2
	2	2	4	2	0	1	1	6	2	0	0	0	6	4	0

An example is given in fig. 24.

Some of these are probably non-existent for one or both forms of boundary.

We have further the contact systems

$$C1,1,1,2, C2,2,1, C2,1,2,$$

which may be left to the reader to examine.

How this geare will cotton I know not.
True Tragedie of Ric. III, 1594.

16. The second set of eight pieces puts up into the parallelogram of fig. 25 with eight boundary compartments.

The set is symmetrical in four colours. Each colour occurs on six compartments. The types for the different contact systems are

	$C_{1,1,1,1}$				$C_{2,1,1}$				$C_{2,2}$		
Colours	1	2	3	4	1	2	3	4	1	2	3	4
	6	2	0	0	4	4	0	0	4	4	0	0
	4	4	0	0	3	3	2	0	3	3	1·1	
	4	2	2	0	2	2	4	0	2	2	0	0
	2	2	2	2	2	2	2	2				
					1	1	6	0				
					1	1	4	2				
					0	0	6	2				
					0	0	4	4				

Examples are given in fig. 25.

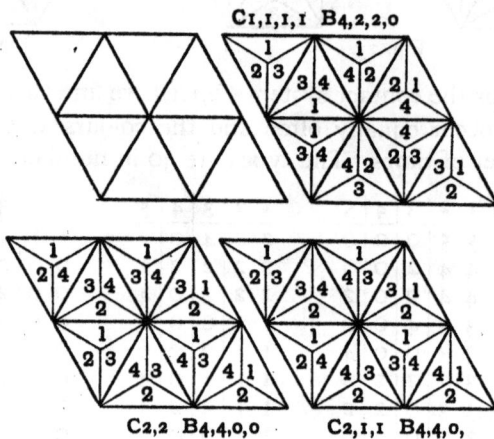

Fig. 25.

SQUARE PASTIMES

As a stream descending
From his fair heads to sea, becomes in trending
More puissant.   G. TOOKE's *Belides*.

17. Taking next the square, divide it into four equal and
similar compartments by drawing the two diagonals and, as
usual, regard each compartment as the location of a colour or
number. Fixing upon three colours and allowing repetitions of
colour in the compartments of the same square we obtain
twenty-four different pieces as in fig. 26 which can obviously be
arranged in a rectangle 6 × 4.

The 24 squares involve 4 × 24 or 96 compartments and, since
each of the three colours is involved symmetrically, each must
appear upon one-third of 96 or 32 compartments. 32 is an even

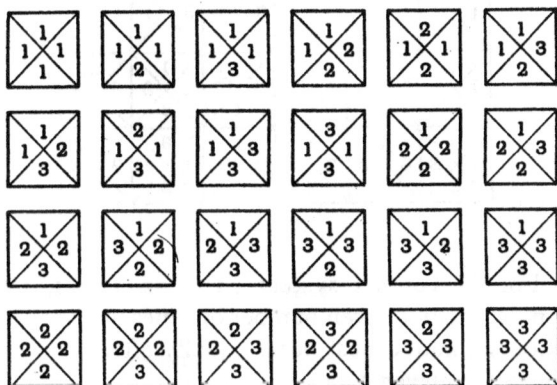

Fig. 26.

number and the rectangle has 20 boundary compartments so
that for the contact system $C_{1,1,1}$, viz. 1 to 1, 2 to 2, 3 to 3,
each colour must appear an even number of times upon the
boundary. We have therefore the 14 types of boundary:

Colours	1	2	3		1	2	3		1	2	3
	20	0	0		14	4	2		10	8	2
	18	2	0		12	8	0		10	6	4
	16	4	0		12	6	2		8	8	4
	16	2	2		12	4	4		8	6	6
	14	6	0		10	10	0				

with varieties of each type except the first.

Most of these types have been actually arranged. Some of them are by no means easy but all are believed to be possible. Very little is known about the varieties of the different types.

The reader will not have much difficulty in setting up the first type $B20,0,0$, with one colour everywhere upon the boundary, but he will find that thought and ingenuity are both required.

For as the precious stone diacletes, though it have many rare and excellent sovraignties in it, yet loseth them all, if it be put into a dead man's mouth. BRAITH, *Engl. Gent.* p. 273.

An arrangement of type $B10,10,0$ is given in fig. 27.

$C1,1,1$ $B10,10,0$

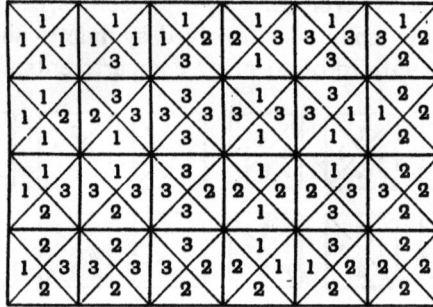

Fig. 27.

18. For the contact system $C1,2$, viz. 1 to 1, 2 to 3, the colour 1 must occur an even number of times upon the boundary, whilst colours 2 and 3 must occur equally often.

We have the eleven types:

Colours	1	2	3		1	2	3
	20	0	0		8	6	6
	18	1	1		6	7	7
	16	2	2		4	8	8
	14	3	3		2	9	9
	12	4	4		0	10	10
	10	5	5				

with the usual varieties. Every type is believed to be a possible arrangement and two examples are given in fig. 28.

Come on, sir frier, picke the locke,
This gere doth cotton hansome.
Troub. Reign of King John, p. 1.

C1,2 B20,0,0

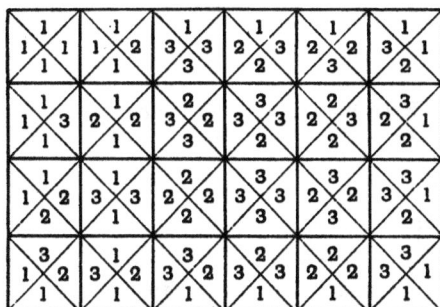

C1,2 B12,4,4
Fig. 28.

Pleased with the same success, vast numbers press'd
Around the shrine and made the same request.

The Temple of Fame.

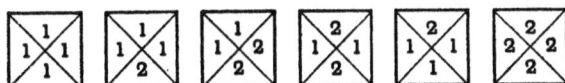

Fig. 29.

19. We may select from the complete set of 24 squares a lesser number upon definite principles.

If we are restricted to two colours we have only six squares as in fig. 29 which fit into a rectangle three squares by two.

The case is trivial from the present point of view, but since transformations of these diversions are effected in Part II of the book it is well not to leave it unexamined.

Briefly the results are that for the contact system $C1,1$ only one boundary type exists, viz. $B6,4$, and for the contact system $C2$ only one, viz. $B5,5$.

Reflection, reason, still the ties improve,
At once extend the interest and the love.

Essay on Man.

20. To make a more useful selection we may in the first place discard all squares which remain unaltered when the colours 2, 3 are interchanged. We thus discard the four squares of fig. 30 and are left with a set of 20 squares which can be assembled into a rectangle 5 × 4.

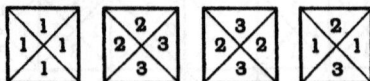

Fig. 30.

The set is not symmetrical in three colours, but only in the colours 2, 3.

There are 18 boundary compartments and the colours 1, 2, 3 occur in the set 26, 27, 27 times respectively.

We see that on the boundary the colours must occur for the system $C1,1,1$

1 an even number of times.
2 „ uneven „
3 „ uneven „

There are 25 boundary types:

Colours 1	2	3	1	2	3	1	2	3
16	1	1	6	7	5	2	11	5
14	3	1	6	9	3	2	13	3
12	3	3	6	11	1	2	15	1
12	5	1	4	7	7	0	9	9
10	5	3	4	9	5	0	11	7
10	7	1	4	11	3	0	13	5
8	5	5	4	13	1	0	15	3
8	7	3	2	9	7	0	17	1
8	9	1						

and the usual varieties.

An example of the type $Bo,9,9$ is given in fig. 31.

$C1,1,1$ $Bo,9,9$

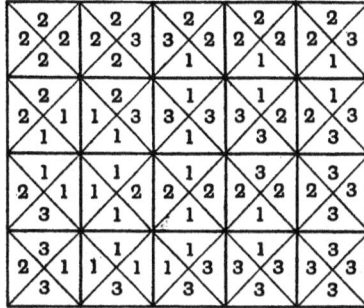

Fig. 31.

21. For the contact system $C2,1$, viz. 1 to 2 and 3 to 3, we find that colour 2 must occur once oftener than 1 upon the boundary and 3 an uneven number of times; this leads to nine types:

Colours	1	2	3		1	2	3
	8	9	1		3	4	11
	7	8	3		2	3	13
	6	7	5		1	2	15
	5	6	7		0	1	17
	4	5	9				

These prove to be difficult to arrange, and some varieties of types appear not to be possible.

Hard or difficile be those thynges that be goodly or honest.
                                        TAVERNER'S *Adagies*, D. 5.

And when he had taryed there a long time for a convenable wind, at length it came about even as he desired.
                                        HOLINSHED'S *Chron.* 1577.

In view of Part II of this book it is desirable to obtain symmetrical boundaries for choice and one such example is given in fig. 32 for the type $B2,3,13$.

22. For the contact system $C1,2$, viz. 1 to 1, 2 to 3, which exists by reason of the want of symmetry in three colours, we

find colour 1 must occur an even number of times upon the
boundary and colour 2 just as often as 3.

C2,1 B2,3,13

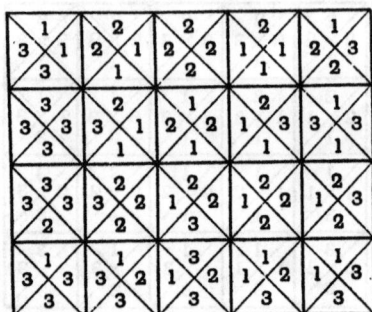

Fig. 32.

There are ten types:

Colours	1	2	3	1	2	3
	18	0	0	8	5	5
	16	1	1	6	6	6
	14	2	2	4	7	7
	12	3	3	2	8	8
	10	4	4	0	9	9

with varieties possibly of all except the first.

As an example a setting of $B18,0,0$ is given in fig. 33.

C1,2 B18,0,0

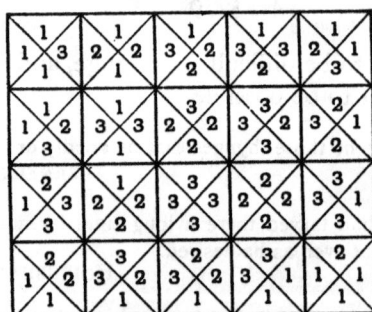

Fig. 33.

23. To reduce further the number of squares and, in the
first place, with the object of obtaining a set of 16 we adopt as a
discarding principle the rejection of all squares which involve a
particular colour exactly twice and upon opposite compartments

of the square. The reasons which prompt this rejection will appear in Part II, as it will be shewn therein that such pieces are inconvenient and even impracticable in the further development. We thus reject the four squares of fig. 34a and we are left with the 16 pieces of fig. 34b.

The set is symmetrical in the two colours 2, 3 but not in the three colours. They form up naturally into a square 4 × 4, which has 16 boundary compartments.

The colours 1, 2, 3 occur upon 20, 22, 22 compartments respectively.

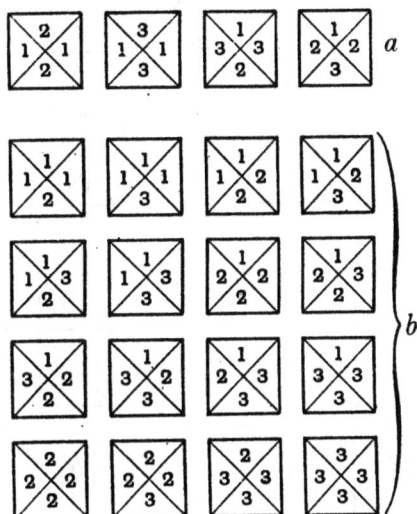

Fig. 34.

For the contact system C1,1,1 each colour must appear an even number of times upon the boundary. We are led to 25 types of boundary:

Colours								
I	2	3	I	2	3	I	2	3
16	0	0	6	10	0	2	10	4
14	2	0	6	8	2	2	8	6
12	4	0	6	6	4	0	16	0
12	2	2	4	12	0	0	14	2
10	6	0	4	10	2	0	12	4
10	4	2	4	8	4	0	10	6
8	8	0	4	6	6	0	8	8
8	6	2	2	14	0			
8	4	4	2	12	2			

with the usual varieties.

It is not known how many of these exist.

> He hath a person, and a smooth dispose
> To be suspected.   *Othello*, I. iii.

As examples we have the three arrangements of fig. 35 in the first of which, $C_{1,1,1}$ $B_{16,0,0}$, symmetry is apparent*.

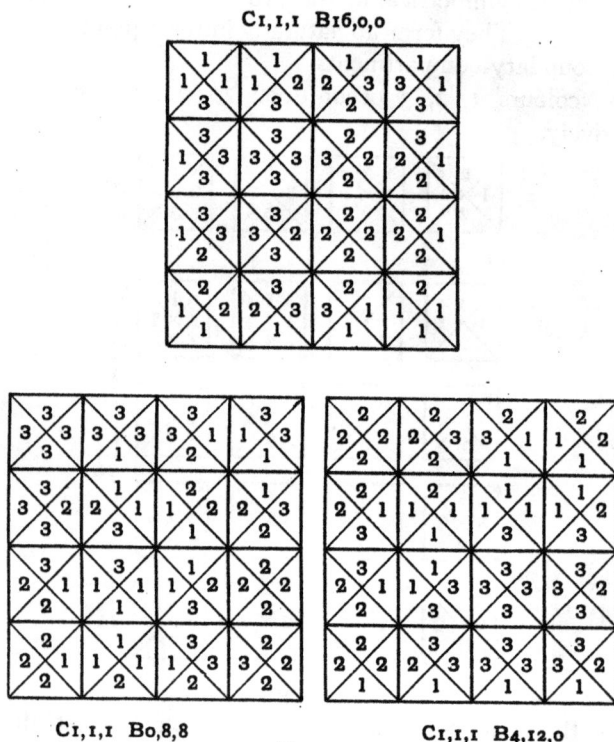

$C_{1,1,1}$ $B_{16,0,0}$

$C_{1,1,1}$ $B_{0,8,8}$ $C_{1,1,1}$ $B_{4,12,0}$

Fig. 35.

24. For the contact system $C_{2,1}$, viz. 1 to 2, 3 to 3, the colour 2 must occur upon the boundary a number of times greater by two than the number of times 1 occurs, while colour 3 must occur an even number of times. The possible types are

Colours 1	2	3		1	2	3
7	9	0		3	5	8
6	8	2		2	4	10
5	7	4		1	3	12
4	6	6		0	2	14

and varieties.

* If any compartment be reflected through the centre of the whole square (the point-image) its colour if 1 is unchanged, but, if 2 or 3, is changed to 3 or 2 respectively.

Examples are given in fig. 36.

C2,1 B0,2,14   C2,1 B2,4,10

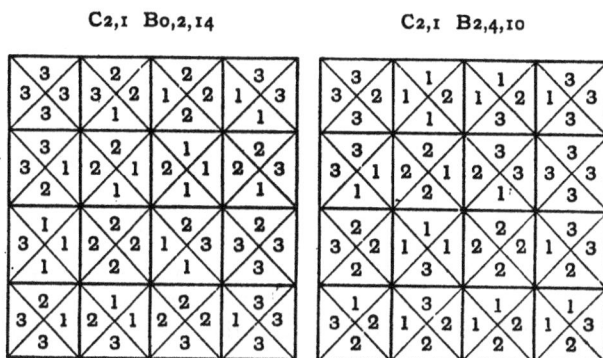

Fig. 36.

25. For the contact system $C1,2$, viz. 1 to 1, 2 to 3, we must have colour 1 an even number of times upon the boundary and colour 2 as often as 3.

The types are nine in number:

Colours	1	2	3		1	2	3
	16	0	0		6	5	5
	14	1	1		4	6	6
	12	2	2		2	7	7
	10	3	3		0	8	8
	8	4	4				

Examples are given in fig. 37.

C1,2 B16,0,0   C1,2 B0,8,8

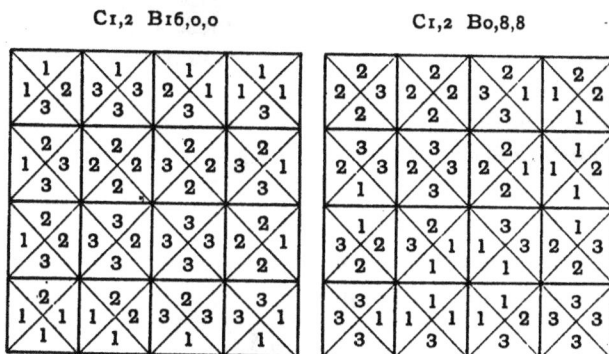

Fig. 37.

This set of pieces has been dealt with at some length because it possesses a large number of boundaries which are symmetrical in colour. Also because it lends itself particularly to the developments of Part II.

> Yes, Sir, I study here the mathematics
> And distillation.
>
> B. JONSON, *Alch.* IV. i.

The numbers 1, 2, 3, ..., which have been used in connexion with the various sets, have been merely symbols for colours.

No use has been made of their arithmetical properties.

A single example will shew that it is possible to denote the colours by numbers and to call in the aid of arithmetic, in certain cases, to define the contact system that is before us.

In the last contact system dealt with above, viz.

> 1 to 1,
> 2 „ 3,

substitute for   the numbers 1, 2, 3,
the numbers 1, 0, 2,

so that the contacts in question are 1, 1 and 0, 2 as in fig. 38*a* and we may define the conditions of contact to be that the sum of the numbers in adjacent compartments is 2. Clearly we

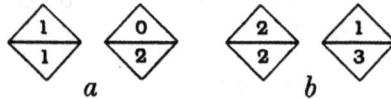

a   *b*

Fig. 38.

could increase each of the numbers 1, 0, 2 by the *same number* with the same result. Thence the simplest way of carrying out the idea is merely to interchange the numbers 1, 2 so that the contacts are 2, 2 and 1, 3 as in fig. 38*b*.

The sum of the numbers in adjacent compartments is then 4.

In the case of five colours 1, 2, 3, 4, 5 and the contact system

> 1 to 1,
> 2 „ 3,
> 4 „ 5,

we similarly change the numbers

> 1, 2, 3, 4, 5

to   3, 1, 5, 2, 4

respectively, and define the contact system to be such that the sum of the numbers in adjacent compartments is 6.

26. There is an interesting set involving 15 pieces which is derived from the set of 16 pieces, dealt with above, by omitting the pieces in fig. 39a and adding the piece in fig. 39b.

This is a 15-piece set which may be defined as involving the pieces of the original set of 24:

(i) which have not more than three compartments of any piece of the same colour;

(ii) which do not involve any piece which has exactly two compartments of the same colour and those opposite compartments.

The pieces will form up into a rectangle 5 × 3 as shewn by the figured rectangle in fig. 39c.

$C_{1,1,1}$ $B_{16,0,0}$

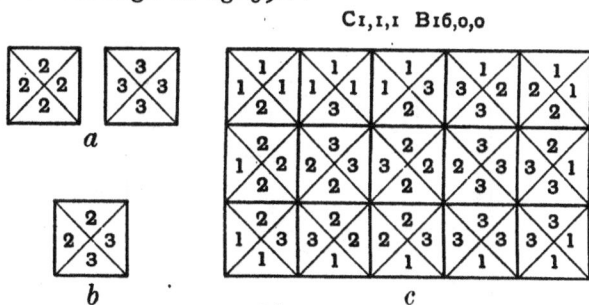

Fig. 39.

To analyse the set we note that it is symmetrical in the three colours and that the compartments coloured 1, 2, 3 occur each 20 times.

There are 16 boundary compartments.

For the contact system $C_{1,1,1}$ each colour must occur an even number of times upon the boundary.

The types are therefore

Colours	I	2	3		I	2	3
	16	0	0		10	4	2
	14	2	0		8	8	0
	12	4	0		8	6	2
	12	2	2		8	4	4
	10	6	0		6	6	4

ten in number.

27. For the contact system $C_{1,2}$, viz. 1 to 1, 2 to 3, the colour 1 must occur an even number of times upon the boundary and the colours 2, 3 equally often.

The types are

Colours 1	2	3	1	2	3
16	0	0	6	5	5
14	1	1	4	6	6
12	2	2	2	7	7
10	3	3	0	8	8
8	4	4			

nine in number.

All the types for both contact systems can probably be arranged and in many varieties.

Above, fig. 39 c, is given an example of the type $B_{16,0,0}$ for the system $C_{1,1,1}$.

In fig. 40 is given one of the type $B_{14,1,1}$ for the system $C_{1,2}$.

$C_{1,2}$ $B_{14,1,1}$

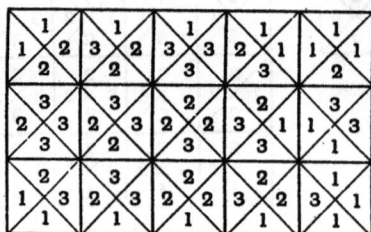

Fig. 40.

28. A 9-piece set can be formed of the six pieces each of which involves three compartments of the same colour and also the three pieces each of which has two colours twice represented on either side of a diagonal. The set is given in fig. 41 assembling into a square 3 × 3.

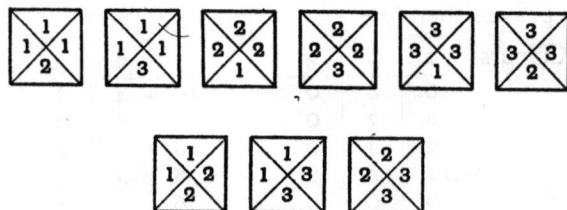

Fig. 41.

The set is symmetrical in three colours, each colour occurring in 12 compartments.

There are 12 boundary compartments.

For the contact system $C_{1,1,1}$, the types are

Colours 1	2	3
8	4	0
8	2	2
6	6	0
6	4	2
4	4	4

five in number.

29. For the contact system $C_{1,2}$, viz. 1 to 1, 2 to 3, the types are

Colours 1	2	3
8	2	2
6	3	3
4	4	4
2	5	5
0	6	6

five in number.

An arrangement of type $B8,2,2$ for the system $C_{1,2}$ is given in fig. 42*a*, the *boundary* colours exhibiting symmetry about the diagonal BD.

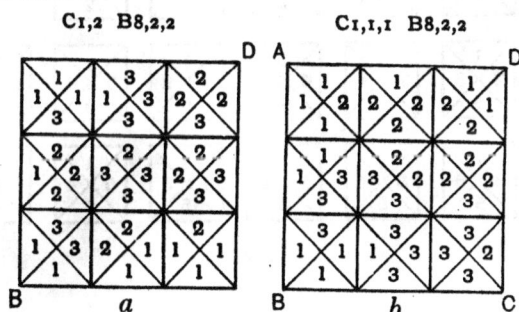

Fig. 42.

One of type $B8,2,2$ for the system $C_{1,1,1}$ is given in fig. 42*b* in which considering the colour symmetry about the diagonal AC, it will be noticed that colour 2 appears along DC in the same way as colour 3 appears along BC.

It is this *kind* of symmetry that leads, for this contact system, to the greatest symmetry in the transformations of Part II.

30. In order to design a set based upon the square and involving five colours we adopt as a principle that every square is to involve four of the colours; there are to be no repetitions of colour in the compartments of the same square.

Four different colours give rise to six squares because four different objects have just six permutations in circular procession. We can select four colours out of five in five different ways so that we can obtain 5 × 6 or 30 different squares involving the five colours. There is little doubt that this would be an interesting set, but in this book we have limited ourselves to sets containing not more than twenty-four pieces and we can reduce the set of 30 as required by importing the condition that every piece is to involve one specified colour, say the colour 1. We thus get the set of twenty-four given in fig. 43 for which the rectangle 6 × 4 is available.

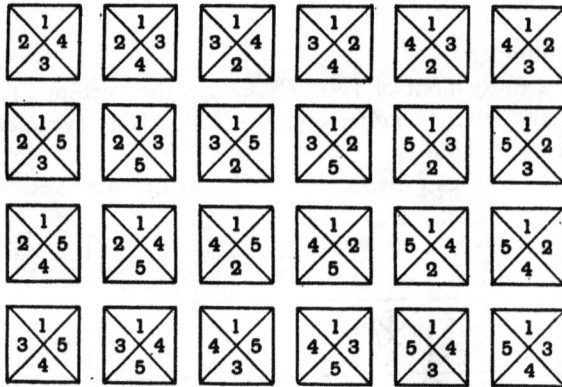

Fig. 43.

The colours 1, 2, 3, 4, 5 are involved upon 24, 18, 18, 18, 18 compartments respectively. These are all even numbers and as the number of boundary compartments is 20, we have as a condition that each colour must appear, in the contact system $C_{1,1,1,1}$, an even number of times upon the boundary. Moreover each angular point of the rectangle must exhibit two different colours, so that a particular colour cannot appear upon the boundary more than 16 times. The number of boundary types is very large; it is not necessary to write them all down. They proceed for the colours

$$
\begin{array}{ccccc}
 & \mathbf{1} & \mathbf{2} & \mathbf{3} & \mathbf{4} & \mathbf{5} \\
\text{from} & 16 & 4 & 0 & 0 & 0 \\
 & \vdots & \vdots & \vdots & \vdots & \vdots \\
\text{to} & 0 & 6 & 6 & 4 & 4
\end{array}
$$

Two examples are given in fig. 44.

C1,1,1,1,1 B14,0,0,4,2

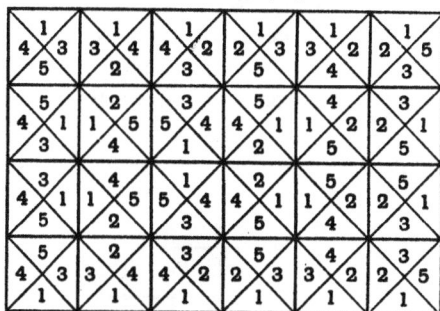

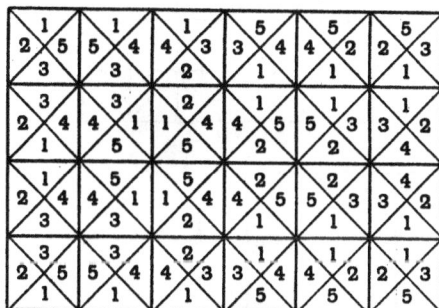

C1,1,1,1,1 B6,6,2,0,6

Fig. 44.

Since there is only symmetry in four colours, four other contact systems are available:

C2,1,1,1	C1,2,1,1	C2,2,1	C1,2,2
1 to 2	1 to 1	1 to 2	1 to 1
3 „ 3	2 „ 3	3 „ 4	2 „ 3
4 „ 4	4 „ 4	5 „ 5	4 „ 5
5 „ 5	5 „ 5		

The reader will know how to set forth the large numbers of boundary types that arise from these systems of contacts.

31. Allusion has been made to varieties of the different types.

These should be chosen, for trial, so as to be symmetrical or semi-symmetrical about some axis passing through the centre of the figure. Thus in the case of the rectangle in fig. 45a the chosen axis may be AB, or CD or any other axis EF passing through the centre of the figure.

If just two equally numerous colours are upon the boundary we may, for the axis AB, take just half of the compartments of each colour on each side of AB and arrange them symmetrically.

Or we may make the compartments to the left of AB any we please and then take a compartment, to the right, of the colour 1 when the corresponding compartment to the left has the colour 2.

Such arrangements for the axes AB, EF might be as in fig. 45b and c.

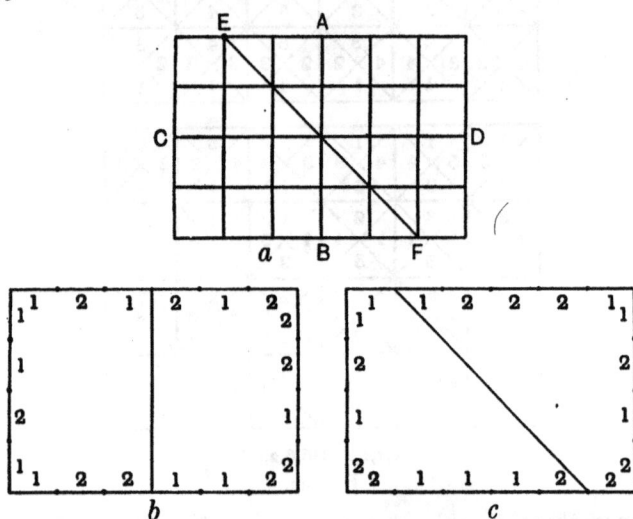

Fig. 45.

These are extreme cases of what should be attempted.

The simplest case would be to place all the compartments of one of the colours on the same side of the axis chosen; and the simplest should always be chosen in the first place. When symmetrical boundaries are not practicable the interest is greatly diminished. It will be found that symmetrical boundaries lend themselves particularly to the transformations of Part II.

RIGHT-ANGLED TRIANGLE PASTIMES

It cottons well, it cannot choose but beare
A pretty napp.

Family of Love, D. 3 b.

32. Take a right-angled triangle, which is half of a square, as in fig. 46*a*.

Find its centre *O* by taking *CO* equal to two-thirds of the perpendicular drawn from *C* upon *AB*. Then the three straight lines *CO, AO, BO* divide the triangle into three compartments of equal areas.

If we have three colours and each triangle is to have three different colours in its compartments we obtain six different triangles which can be assembled, with the contact system *C*1,1,1, in the form of fig. 46*b* with any chosen colour monopolising the boundary.

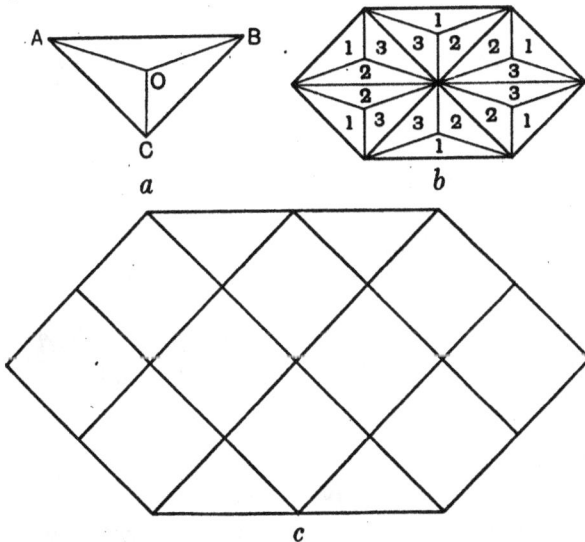

a *b*

c
Fig. 46.

If we are given four colours we are led to 24 different triangles which can be arranged in the hexagonal form of fig. 46*c*, a figure consisting of four triangles and ten squares.

Two triangles may fit into any one of these squares with the long sides either vertical or horizontal. This is a new feature which adds interest to the study of the pieces.

This is exhibited by two arrangements for the contact system $C1,1,1,1$ which are given in fig. 47.

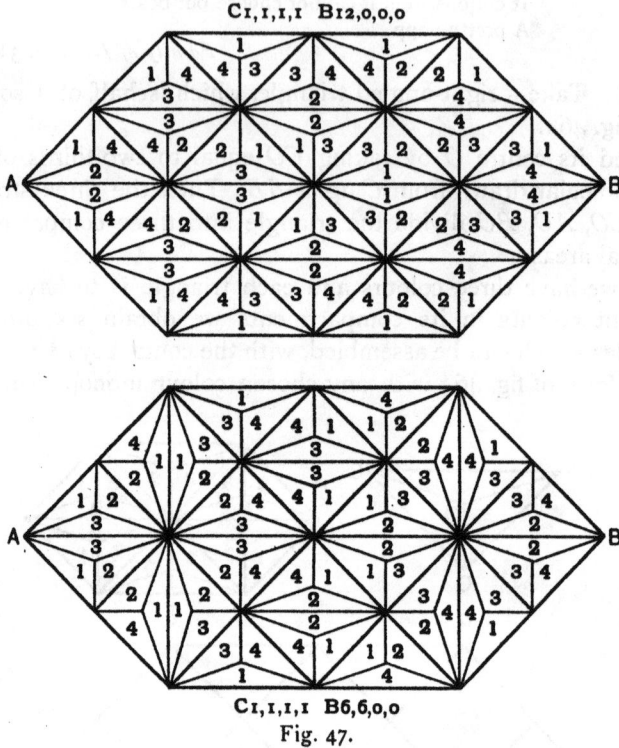

C1,1,1,1 B12,0,0,0

C1,1,1,1 B6,6,0,0
Fig. 47.

The first is symmetrical in colour about the axis AB.

The second differs from the first both in type of boundary and in internal structure.

In the first the triangles have two aspects, and in the second four aspects. In order to deal with the internal structure by type we may suppose the type to depend upon the numbers of triangles which have their long sides horizontal and vertical respectively. We might then describe the first arrangement by

$$B12,0,0,0 \quad A24,0,$$

and the second by

$$B6,6,0,0 \quad A16,8.$$

Altogether there are 72 compartments, 24 long (L for long) and 48 short (S for short). Since there is symmetry in four colours, each colour appears upon $6L$ and $12S$ compartments.

Upon the boundary there are $4L$ and $8S$. Hence for the contact system $C1,1,1,1$, each colour must have an even number of L and an even number of S compartments upon the boundary.

So that as regards L we have for

Colours	1	2	3	4	
	4	0	0	0	
	2	2	0	0	two types;

and as regards S we have for

Colours	1	2	3	4	
	8	0	0	0	
	6	2	0	0	
	4	4	0	0	
	4	2	2	0	
	2	2	2	2	five types;

with numerous varieties both in internal structure and in boundary.

C1,1,2 B12,0,0,0 A24,0

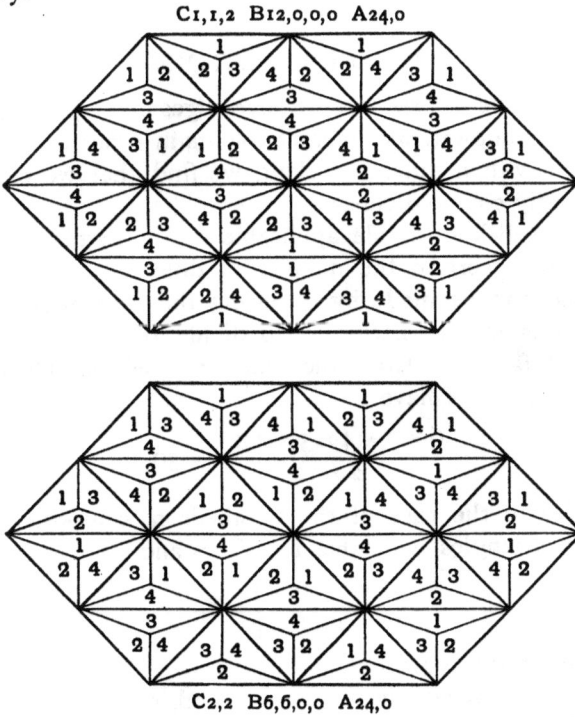

C2,2 B6,6,0,0 A24,0

Fig. 48.

We must presume that each of the L types may occur with each of the S so that there *may* be ten types in all.

Also of internal structure we *may* have the eleven types

A24,0, A22,2, A20,4, A18,6, A16,8, A14,10, A12,12,

A10,14, A8,16, A6,18, A4,20,

and of every type of boundary except one and of every type of internal structure except one there *may* be varieties.

In regard to other contact systems, there is no reason why the L compartments should have the same system as the S, because they do not clash at all.

33. Other systems available for both L and S are

C1,1,2, viz. 1 to 1, 2 to 2, 3 to 4,

and C2,2, viz. 1 to 2, 3 to 4.

Assemblages for the systems (for *both* L and S) C1,1,2, C2,2 are given in fig. 48.

THE CUBE PASTIME

34. A cube has six faces, twelve edges and eight summits.

If we are allowed six different colours in order to colour the faces each with a different colour, we find that we can make 30 differently coloured cubes.

It is a well-known rule, applicable to any regular solid, that in order to ascertain the number of different cubes or other solids that can be made by colouring the faces with different colours it is merely necessary to divide the factorial of the number of faces by twice the number of edges. Thus in the case of the cube we have

$$\frac{6 \times 5 \times 4 \times 3 \times 2 \times 1}{2 \times 12} = 30.$$

So also in the case of the tetrahedron, composed of four equilateral triangles, which has four faces and six edges we have

$$\frac{4 \times 3 \times 2 \times 1}{2 \times 6} = 2,$$

and so on for any regular solid.

We now construct these 30 cubes and, denoting the colours by numbers, we represent any such cube in a diagram as in fig. 49 a, the cube being supposed resting upon a table and viewed from above.

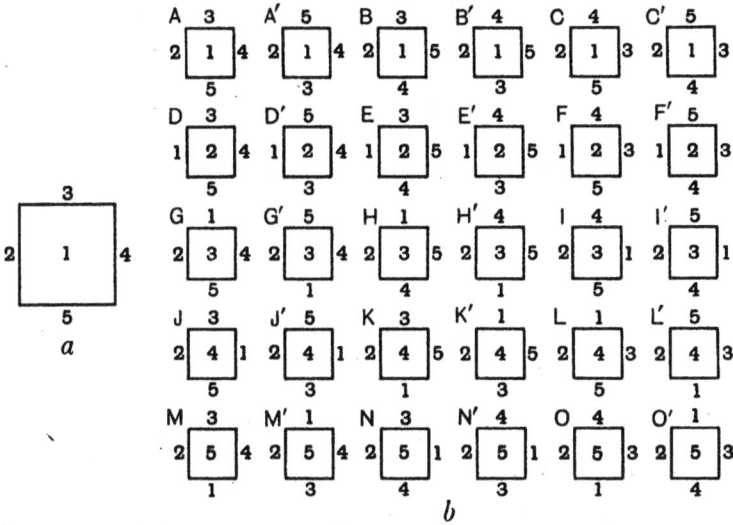

Fig. 49 b (the 30 cubes). Each cube shows its top number, left number, centre (uppermost face = 1, etc.), right number, and bottom number:

A 3 2[1]4 5	A′ 5 2[1]4 3	B 3 2[1]5 4	B′ 4 2[1]5 3	C 4 2[1]3 5	C′ 5 2[1]3 4
D 3 1[2]4 5	D′ 5 1[2]4 3	E 3 1[2]5 4	E′ 4 1[2]5 3	F 4 1[2]3 5	F′ 5 1[2]3 4
G 1 2[3]4 5	G′ 5 2[3]4 1	H 1 2[3]5 4	H′ 4 2[3]5 1	I 4 2[3]1 5	I′ 5 2[3]1 4
J 3 2[4]1 5	J′ 5 2[4]1 3	K 3 2[4]5 1	K′ 1 2[4]5 3	L 1 2[4]3 5	L′ 5 2[4]3 1
M 3 2[5]4 1	M′ 1 2[5]4 3	N 3 2[5]1 4	N′ 4 2[5]1 3	O 4 2[5]3 1	O′ 1 2[5]3 4

Standalone cube a:
3
2[1]4
5

b

Fig. 49.

Thus we are looking straight down upon the uppermost face coloured 1 and see the edges of the four vertical faces which are numbered 2, 3, 4, 5 respectively: the faces are read always in clockwise order.

The face numbered 6 is not visible, so that the number 6 does not appear in the diagram but is regarded as being present but obscured from view by the mass of the cube.

This being understood the 30 cubes are as given in fig. 49 b.

They are conveniently denoted by 15 capital letters and by the same letters dashed, because they naturally arrange themselves into 15 pairs of cubes.

For example the cubes denoted by G, G′ have the same pairs of opposite faces; the faces coloured 1, 2, 3 are opposite to those coloured 5, 4, 6 respectively in both cubes. If the colours upon any pair of opposite faces of one of the cubes be interchanged the other cube is produced.

Looking vertically down upon the cubes the colours read clockwise on the one are identical with the colours read counterclockwise upon the other; the readings having reference of course to the vertical faces.

The above cubes are called 'associated cubes.'

It is not obvious or even very easy to construct a Pastime from these 30 cubes. They can be assembled into a block having the dimensions $2 \times 3 \times 5$ and we can make selections from the whole number in many ways; for instance if we can select intelligently either 8 or 27 of these they can be assembled into large cubes. Moreover we have four different contact systems at our disposal which, following the practice of other pastimes, we might denote by $C1,1,1,1,1,1$, $C1,1,1,1,2$, $C1,1,2,2$, $C2,2,2$.

It is now some years since Colonel Julian R. Jocelyn communicated to the present writer the fact that he could select eight cubes and assemble them on the contact system $C1,1,1,1,1,1$, that is,

$$1 \text{ to } 1, 2 \text{ to } 2, 3 \text{ to } 3, 4 \text{ to } 4, 5 \text{ to } 5, 6 \text{ to } 6,$$

so as to produce a cube of twice the linear dimensions which is a faithful copy in colours of any *given* member of the set of thirty cubes.

Suppose that it is desired to thus produce the cube denoted by A.

The *two* cubes A, A' have the opposite pairs 1—6, 2—4, 3—5.

Reject from the complete set all the cubes which have any pair of these opposites and it will be found that we are left with the following 16, viz.:

$$E, F, H, I, K, L, N, O,$$
$$E', F', H', I', K', L', N', O'.$$

These may be further divided into two sets, each of eight cubes, viz.:

first set $K, L, F', E', H', O', I, N,$
second set $K', L', F, E, H, O, I', N',$

which are connected with the cubes A, A' respectively.

The first set may be assembled, with the contact system $C1,1,1,1,1,1$, in two distinct ways so as to reproduce the cube A on twice the linear scale. The first solution is shewn in fig. 50a, the second solution in fig. 50b, the result, in each case, being as in fig. 50c.

The two solutions are connected in an interesting manner. A straight line which passes from any summit of a cube through the centre of the cube to another summit is a *diagonal* of the cube. There are four diagonals. Each of the eight cubes which compose the large cube is *diagonally* opposite to another cube.

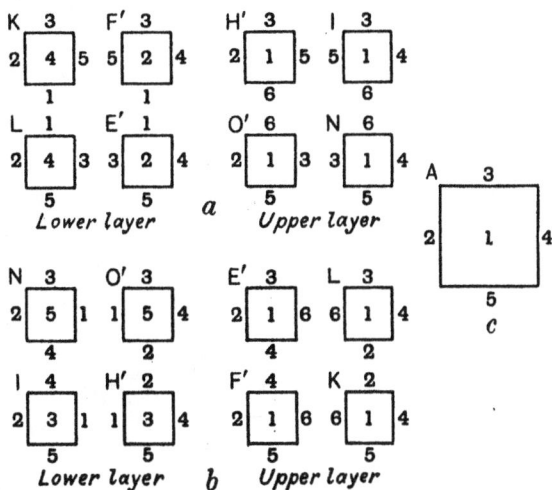

Fig. 50.

We see that in each solution there is *no* change in the diagonally opposite pairs. In the first solution, for example, we have the pair $F'O'$ corresponding to the pair $O'F'$ in the second. All that has happened is that the components of the pair have become interchanged.

The eight cubes in either case involve 48 faces and of these exactly half, viz. 24, are boundary faces. The remaining 24 are inside faces. It is a remarkable circumstance that the 24 boundary faces in the first solution are inside faces in the second and *vice versa*.

The geometry of the solutions can be further studied by taking advantage of the fact that the six centres of the six faces of a cube are the summits of a regular octahedron*. The geometrical reader may be interested in following up this point.

In order to make an interesting property of the set of 30 cubes clear, it is convenient to speak of the large cube A as *containing* each of the component cubes $K, L, F', E', H', O', I, N$.

* *Proceedings of the London Mathematical Society*, Vol. 24, p. 145, 1893.

In this language the complete results may be set forth as follows :

A	contains	*KLF'E'H'O'I'V*	*A'*	contains	*FEK'L'I'N'HO*
B	„	*MO'FD'G'LI'J*	*B'*	„	*F'DM'OIJ'GL'*
C	„	*H'GD'EKM'JN'*	*C'*	„	*DE'HG'J'NK'M*
D	„	*LKC'B'GMI'N'*	*D'*	„	*CBL'K'ING'M'*
E	„	*O'MCA'HKIJ'*	*E'*	„	*C'AOM'I'JH'K'*
F	„	*GH'A'BLOJ'N*	*F'*	„	*AB'G'HJN'L'O'*
G	„	*CB'O'NFK'DJ*	*G'*	„	*ON'C'BD'J'F'K*
H	„	*C'A'LJF'M'EN*	*H'*	„	*L'J'CAE'N'FM*
I	„	*B'AED'K'MLO*	*I'*	„	*E'DBA'L'O'KM'*
J	„	*BCE'F'HOGM*	*J'*	„	*EFB'C'G'M'H'O'*
K	„	*GI'ACENDO*	*K'*	„	*A'C'GID'O'E'N'*
L	„	*N'M'BAIFHD*	*L'*	„	*B'A'NMH'D'I'F'*
M	„	*JL'DEIC'H'B*	*M'*	„	*D'E J'LHBI'C*
N	„	*AC'D'FGL'HK*	*N'*	„	*DF'A'CH'K'G L*
O	„	*A'B'JKG'E'IF*	*O'*	„	*J'K'ABI'F'GE*

The property for observation is that if any cube *X* contains the cube *Y*, then reciprocally the cube *Y* contains the cube *X*.

For example *K* being in the set of eight which compose *A*, it will be found on inspection of the results that *A* is in the set which compose *K*.

The first result, which specifies the cubes contained by *A*, shews at once the eight cubes, each of which contains *A*.

The geometry of the octahedron referred to above reveals the reason for this reciprocity.

It is probable that much more remains to be discovered concerning the properties of the set of 30 cubes. The writer has no doubt that clever readers will find out other selections and contact systems which have escaped his own observation.

As a Pastime it is not difficult to assemble any given set of eight cubes if the cube to be assembled is known beforehand ; but it is not nearly so easy if this knowledge be withheld from the solver. It is not however, in that case, a mere matter of chance ; there is ample opportunity, as in all of these Pastimes, for the exercise of thought and cleverness.

35. The circumstance, that it is possible to select eight cubes from the complete set which, on a defined contact system, reproduce on a larger scale a particular cube selected from the complete set, has its parallel in the Triangle and Square Pastimes in the earlier pages of this book. Thus if we take the set of 20 equilateral triangles associated with five colours, not repeatable, we find that four triangles may be selected which will reproduce on twice the scale any chosen member of the set. Not only so, but nine triangles may be selected so as to reproduce it on three times the scale and sixteen triangles may be chosen so as to reproduce it on four times the scale, as in fig. 51 where arrangement *a* reproduces triangle *b* on four times the scale.

Also in the case of the set of 24 right-angled triangles connected with four colours, not repeatable, we can select four triangles with the same result as in fig. 51 *c* and *d*. The same property is possessed by the squares associated with non-repeatable colours as shewn in fig. 51 *e* and *f.*

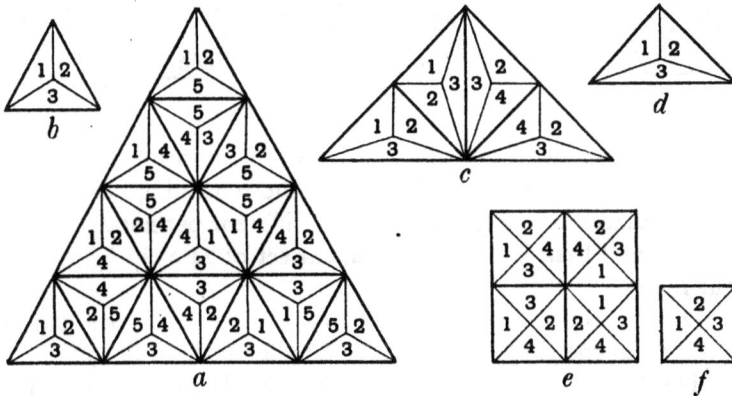

Fig. 51.

PASTIMES BASED UPON THE REGULAR HEXAGON

As our degrees are in order distant,
So the degrees of our strengths are discrepant.
HEYWOOD'S *Spider and Flie*, 1556.

36. The compartments of the hexagon in the first row of fig. 52 are numbered in circular order. We must define some law with regard to the numbers, or to the colours which they may denote, which leads to a convenient number of hexagons. If we have six different colours upon each hexagon we are led to a set of 120 because, placing a particular colour, say 6, uppermost, we can permute the remaining colours in $5 \times 4 \times 3 \times 2 \times 1$ different ways.

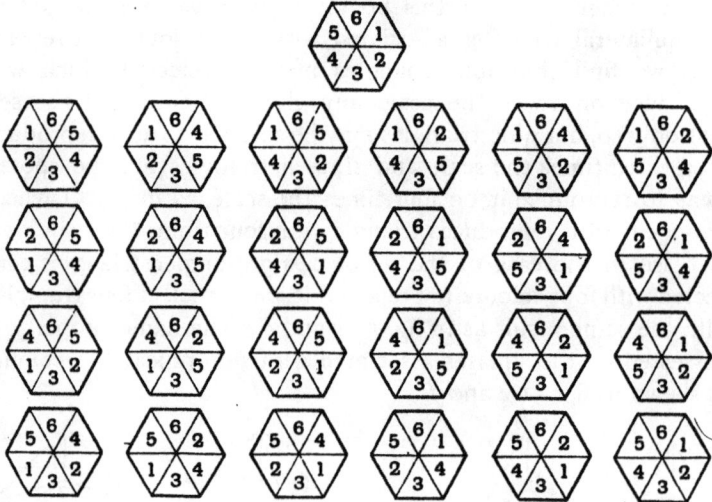

Fig. 52.

If however we not only place the colour 6 uppermost but also another colour, say 3, downmost, the remaining colours can only be placed in $4 \times 3 \times 2 \times 1$ or 24 different permutations.

We shall then have the convenient set of 24 as in fig. 52 (last four rows).

On the contact system $C_{1,1,1,1,1,1}$ these hexagons can be arranged in the shapes shewn on a smaller scale in fig. 53.

The forms of assemblage marked I, II, III have 38, 40 and 36 boundary compartments respectively.

The set of 24 hexagons involves the compartments coloured 1, 2, 3, 4, 5, 6, each 24 times. It follows that for the contact system $C_{1,1,1,1,1,1}$ each colour must occur an even number of times upon the boundary in each of the three assemblages.

The two colours denoted by 3, 6 are in an exceptional position because they are opposite upon each piece.

If we consider any piece in either of the three assemblages and join the centres of its 3, 6 compartments, this line produced both ways will always pass through 3 and 6 compartments (alternately) and through no other until the two opposite boundaries are reached. Further, any straight line joining the centres of 3 and 6 compartments of any other piece in the assemblage must be parallel to the foregoing line.

It follows at once that, since the boundary is made up of lines going in three different directions, one of these directions must involve the colours 3 and 6 exclusively.

We examine the boundaries of the assemblages I, III and discover that the colours 3, 6 together must occupy either 16 or

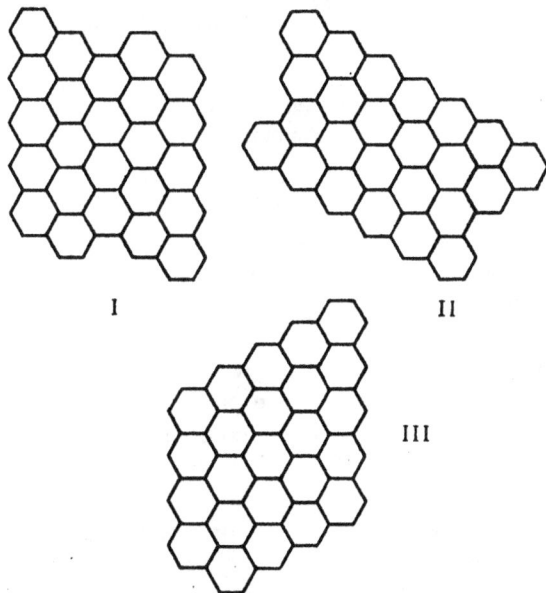

Fig. 53.

10 boundary compartments and each must occur not less than four times. In the case of assemblage II the colours 3, 6 together must occupy either 8 or 16 boundary compartments and each must occur not less than twice. We have therefore types depending upon these two colours as follows:

Colours	I				II				III			
	3	6	3	6	3	6	3	6	3	6	3	6
	12	4	4	12	14	2	4	12	12	4	4	12
	10	6	6	4	12	4	2	14	10	6	6	4
	8	8	4	6	10	6	6	2	8	8	4	6
	6	10			8	8	4	4	6	10		
					6	10	2	6				

whilst, in each of the three assemblages, the colours 1, 2, 4, 5 must each occur an even number of times upon the boundary.

PART II

THE TRANSFORMATION OF PART I

Kind neighbours; mutual amity prevails;
Sweet interchange of rays, received, return'd;
Enlight'ning, and enlighten'd! All, at once,
Attracting, and attracted! Patriot-like,
None sins against the welfare of the whole;
But their reciprocal, unselfish aid
Affords an emblem of millennial love.   *The Consolation.*

37. In Part I we have had before us sets of triangles, squares, etc., which, as regards a particular set, are all of the same size and shape, but are differently coloured or numbered; and we have seen that a particular set of pieces may be set up into a square, rectangular, hexagonal or other shape so that certain contact laws inside the boundary of the figure are satisfied. For a particular set we have an associated outline inside of which the whole of the pieces are arranged. The form of the boundary of the figure is composed of straight lines and does not alter however the relative positions of the pieces may change. Although the geometry of the figure does not vary, the boundary is differently coloured or numbered for each type of boundary and for each variety of type.

The question now before us is the transformation of the set of pieces so that they will no longer be of the same shape and differently coloured or numbered. It proves to be possible to effect this so that the pieces are of different shapes, and are not differently coloured, numbered or otherwise distinguished. In fact, instead of having a set of pieces of the same shape and differently coloured we construct a set of different shapes but of the same colour.

The boundary of the assembled pieces now varies in shape with each type and variety but is not otherwise distinguished.

Studia prima la scienza, e poi seguita la pratica nata da essa scienza.
                    LEONARDO DA VINCI.

Suppose that we have two equilateral triangular pieces as in fig. 54 a and that the contact system involves a compartment of colour 1 being adjacent to a compartment of the same colour.

At present the boundary of the compartment concerned is the straight line *AB*. Is it possible to substitute another boundary which will still allow the two compartment boundaries to lie up against one another?

In fig. 54 *b* bisect *AB* in *E* and *OB* in *D*. Draw *AF* at right angles to *AC*. Join *DE* and produce to meet *AF* in *F*.

Take *AFDB* as a new compartment boundary in both triangles and it will be found that by rotations of the left-hand triangle clockwise about *B* and of the right-hand one counter-clockwise about *A*, the two new boundaries will lie perfectly up against one another, as shewn in fig. 54 *c* where *AFDB* is the common boundary of pieces which are still of the same size and shape.

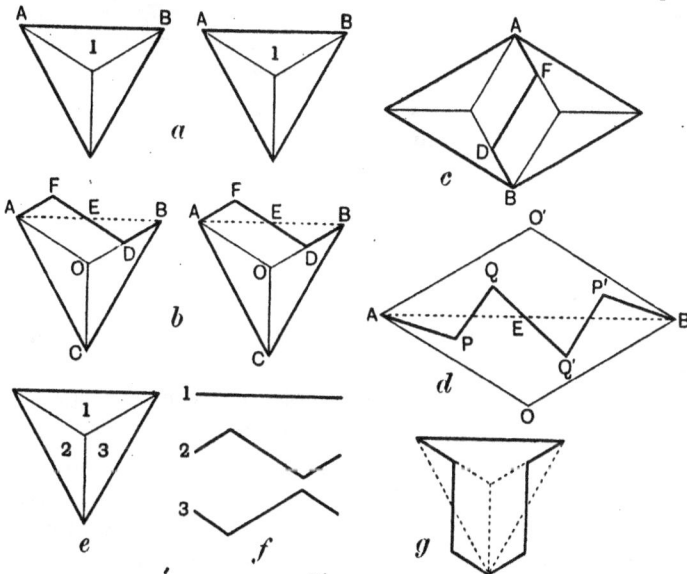

Fig. 54.

A compartment originally coloured 1 will always fit as desired after the transformation provided that no compartment of a different colour is transformed in the same way.

It will be noticed that the transformation does not alter the area of the piece and that the colour 1 is no longer necessary. There is no limit to the number of ways in which the alteration of a compartment boundary can be carried out.

Consider the parallelogram *AOBO'* in fig. 54 *d* formed by two adjacent compartments and *any* point *P* inside; also the corresponding point P' found by joining PE and producing it an equal distance to P'.

Also select any other point Q inside and find the corresponding point Q'. Then taking $APQQ'P'B$ as the common boundary of the two pieces we see that the two pieces will be of the same size and shape and will fit into one another perfectly.

Any number of such points P, Q may be taken either upon the boundary of the parallelogram or inside of it, and an appropriate compartment boundary will result.

In particular the points may be infinite in number so that a curve is produced by joining them.

It follows that the transformed boundary may be determined in infinite variety.

Suppose then that we have any number of colours and the first contact system in which each compartment is adjacent to one similarly coloured, we are able to select any convenient boundaries to associate with the colours and *one* of these may be the straight line, the original compartment boundary. The shape of each compartment of each piece may thus be transformed and the pieces themselves given various curious shapes depending upon the choice of compartment boundaries.

The boundary shapes that are found to be the most suitable for the equilateral triangle are not necessarily the best for the square, hexagon, etc.

To give in this place one example, suppose that fig. 54 e is one of a set of triangles, we may associate with the colours the boundaries as in fig. 54 f transforming the piece to the colourless shape of fig. 54 g.

If, in the coloured or untransformed set of triangles, the type be such that the colour 1 is to monopolise the boundary, it will not be changed because the boundary associated with the colour 1 has not been altered. But if the type of boundary be other than that which has been specified the boundary of the setting will exhibit some of the new compartment boundaries that have been introduced. Thus the pieces instead of setting up always into the same shape, differently coloured or marked, set up into different colourless shapes.

This circumstance adds much to the interest of the pastime of fitting the pieces into shapes which have been previously determined, by the methods of Part I, to be possible.

38. In the case of the second kind of contact we have, say, a colour 1 lying invariably up against a colour 2 and, whatever may be the number of colours, we must arrange such a transformation of compartment boundaries that the colour 1 *will* lie up against colour 2 but will *not* lie up against any of the colours 1, 3, 4, 5, ...

Taking, as before, the equilateral triangle as an illustration, we must arrange for the compartment boundary of colour 1 to fit into the boundary we associate with colour 2, taking care that the two boundaries differ from one another.

If to compartment 1 we add on a projecting piece *AKB* (fig. 55 *a*), equal and similar to *AOB* so that the angle *KAC* is a right angle, and cut out the compartment 2 altogether, it is clear that the projecting piece *AKB* will exactly fit into the recess *AJB*, and that the combined area of the transformed triangles will not be changed.

The figure 55 *b* ‚shews the piece *CAKB* fitting the piece *AKBC'*. The conditions are satisfied because the compartment boundaries *AKB*, *AJB* are quite different.

We must now discover the most general transformation of this nature. It is clear that the area of each triangle (but not the combined area) will usually be altered by the transformation.

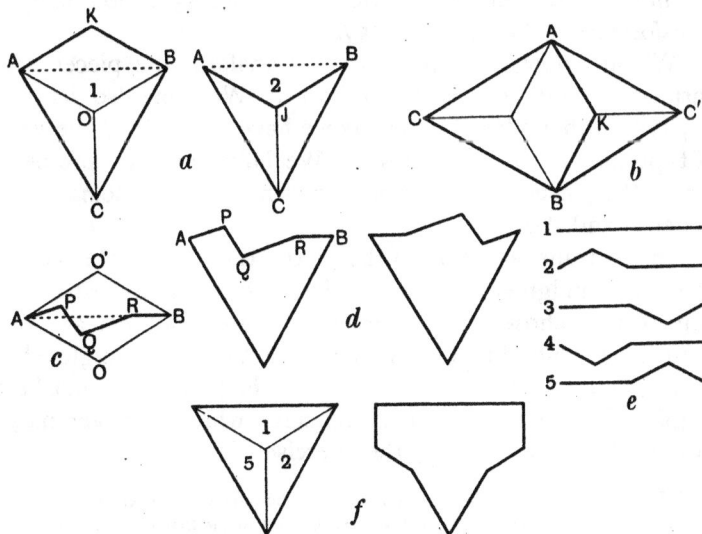

Fig. 55.

As will appear, the area of each triangle is only exceptionally unaltered.

Take, as before, the parallelogram of the compartments $OAO'B$, fig. 55 c. We can select in the area any number of points we please, which may lie if we like either upon the boundaries of the parallelogram or upon AB, and join them by straight lines. We shall then have suitable boundaries to the compartments as shewn in fig. 55 d.

A necessary precaution is to avoid the kind of symmetry about the middle point of AB that always exists in transformations connected with the first kind of contact.

When the number of points selected $P, Q, R, \ldots$ is infinite we are led to boundaries wholly or partially curved.

We take separate pairs of compartment boundaries for the different associated pairs of colours in the contact system.

Each compartment of each triangle is thus transformed *.

As an example, suppose that we have five colours and that the contact system is $C_{1,2,2}$:

1 adjacent to 1

2   „   3

4   „   5

we may select the boundaries as in fig. 55 e, obtaining the transformation shewn in fig. 55 f.

We are now in a position to transform the pieces of the various pastimes described in Part I. We can also transform any type of boundary as soon as we have determined the variety of type that is to be employed. We have then the problem of fitting the pieces into the boundary, which usually demands both patience and thought.

In the transformation of the pieces there is ample scope for taste and judgment. The simplest straight-line forms should generally be chosen as it is easier to construct them accurately than those which involve circular arcs. Very acute angles should be avoided as far as possible so that the pieces may not be too fragile. This presents a real difficulty which however may be practically surmounted by the exercise of ingenuity.

---

* It is not essential for the compartment boundary to lie entirely within the parallelogram of compartments in this or in the case of the first kind of contact. The condition however is in most cases advisable and may be recommended.

Quel che è nuovo è sempre bello.

GOLDONI.

39. The transformation of the 24 pieces of Pastime no. 2, four colours on a triangle with repetitions allowed, for the contact system $C_{1,1,1,1}$ may be carried out in the following manner. Taking the boundaries as in fig. 56 we get the pieces as in fig. 57. In fig. 58 these are shewn assembled inside a boundary of regular hexagonal shape.

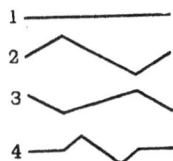

Fig. 56.

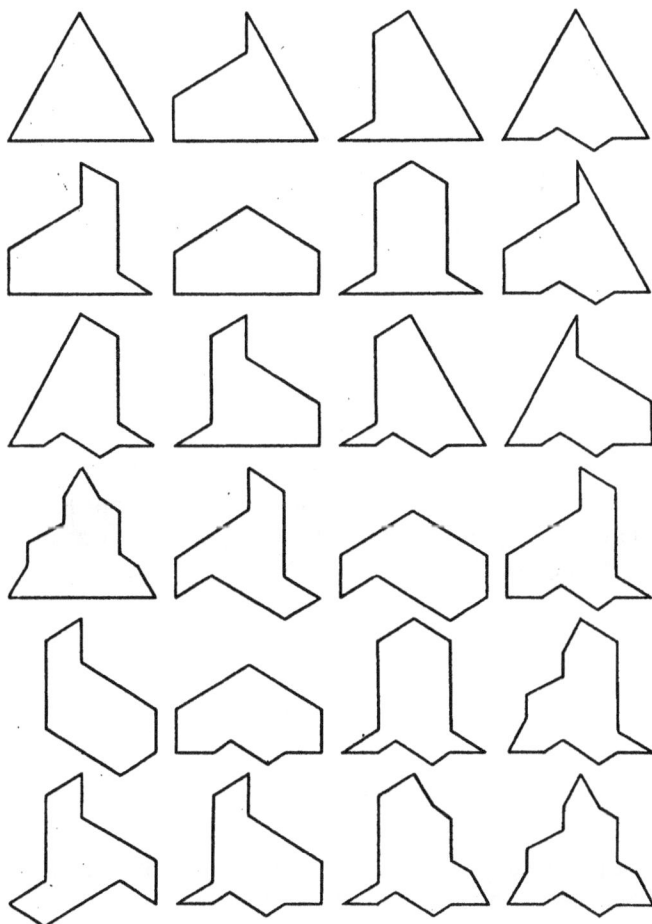

Fig. 57.

> Nor think thou seest a wild disorder here :
> Through this illustrious chaos to the sight,
> Arrangement neat, and chastest order reign.
> The path prescribed, inviolably kept
> Upbraids the lawless sallies of mankind.

The Consolation.

$C_{1,1,1}$ $B_{12,0,0,0}$

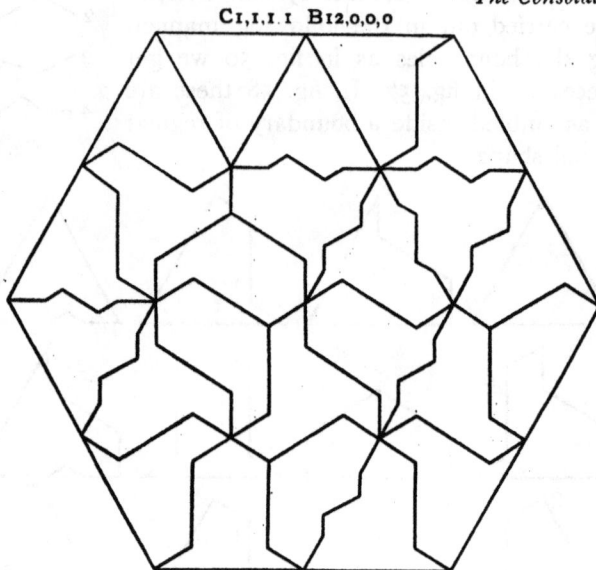

Fig. 58.

Nil fuit unquam
Sic impar sibi.   HORACE.

40. For each contact system transformation leads to a different set of pieces. Thus for the system $C_{1,1,2}$ or 1 to 1, 2 to 2, 3 to 4, we choose the boundaries as in fig. 59.

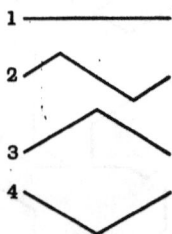

Fig. 59.

The piece each of whose compartments is coloured 4 vanishes with the transformation and we obtain the twenty-three pieces shewn in fig. 60.

Bring then these blessings to a strict account;
Make fair deductions; see to what they 'mount':
How much of other each is sure to cost;
How each for other oft is wholly lost.

POPE, Ess. Man, IV. 270.

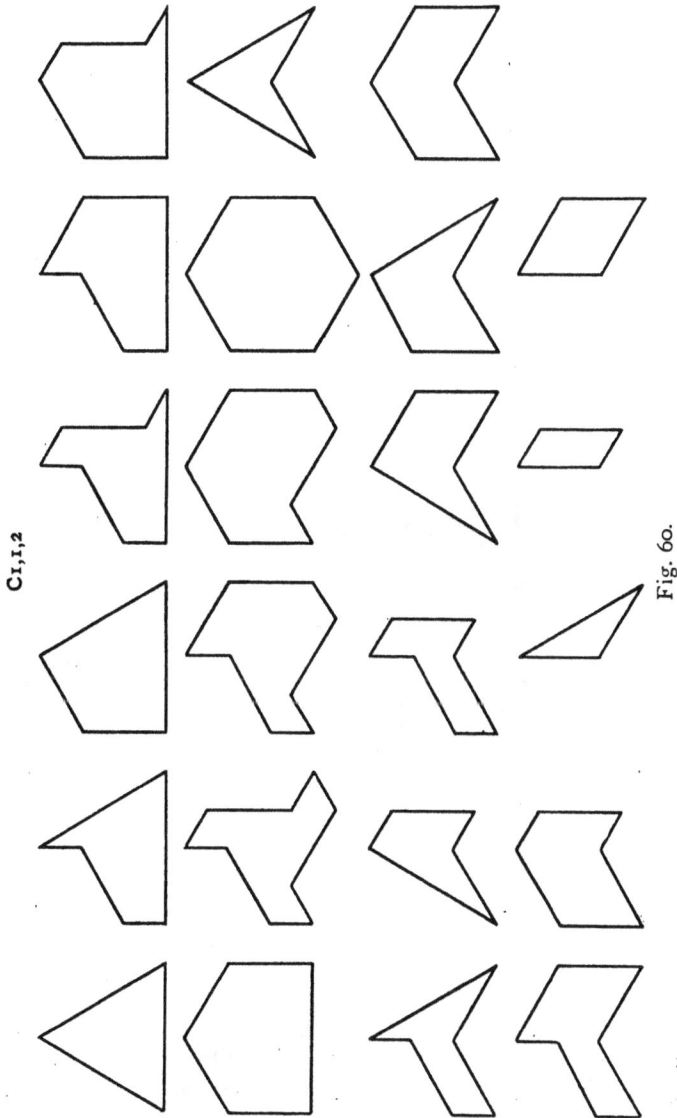

Fig. 60.

$C_{I,I,2}$

The assemblage of the pieces according to fig. 61, which is the figured diagram of Part I, is shewn in fig. 62.

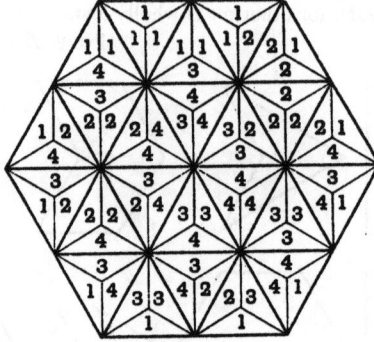

Fig. 61.

C1,1,2

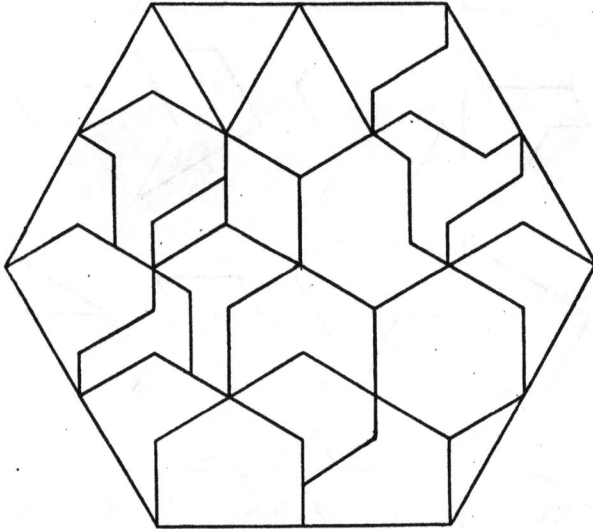

Fig. 62.

Sic ludus animo debet aliquando dari,
Ad cogitandum melior ut redeat tibi.

PHAEDR. *Fab.* III. 14.

41. For the third contact system $C2,2$ which is 1 to 2, 3 to 4 we select for the colours as in fig. 63; one piece again vanishes for the transformation and we are left with the twenty-three pieces shewn in fig. 64.

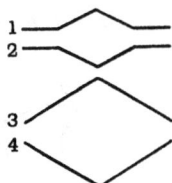

Fig. 63.

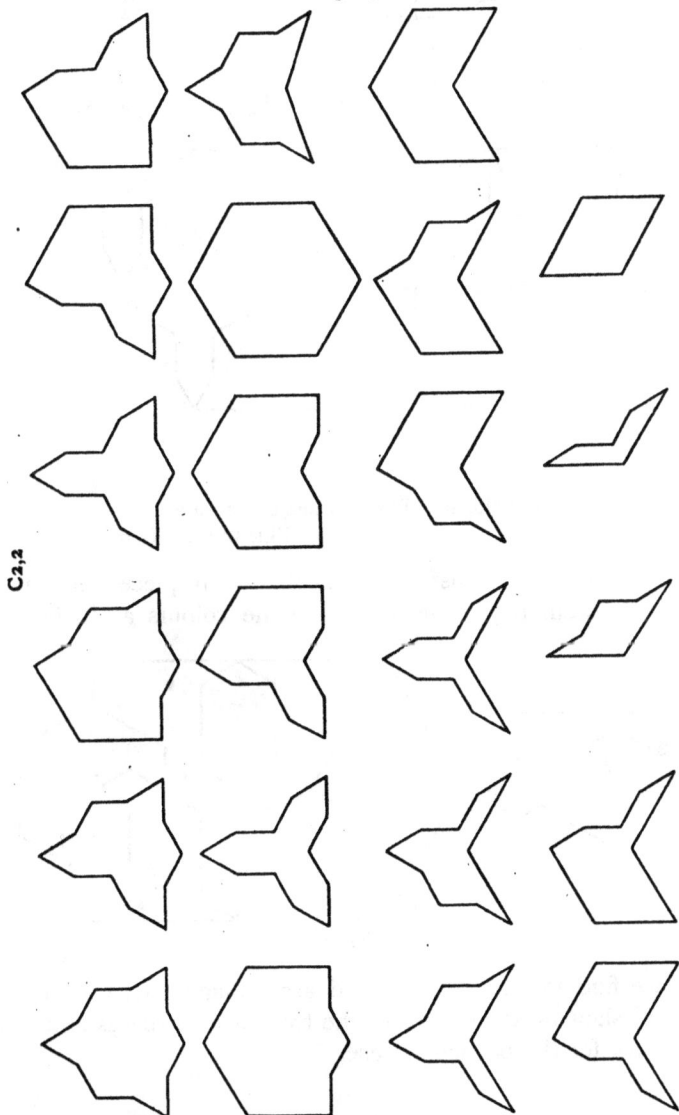

C$_{2,2}$

Fig. 64.

In fig. 65 the pieces are assembled according to the diagram $C2,2$ $B6,6,0,0$ of Part I.

C2,2

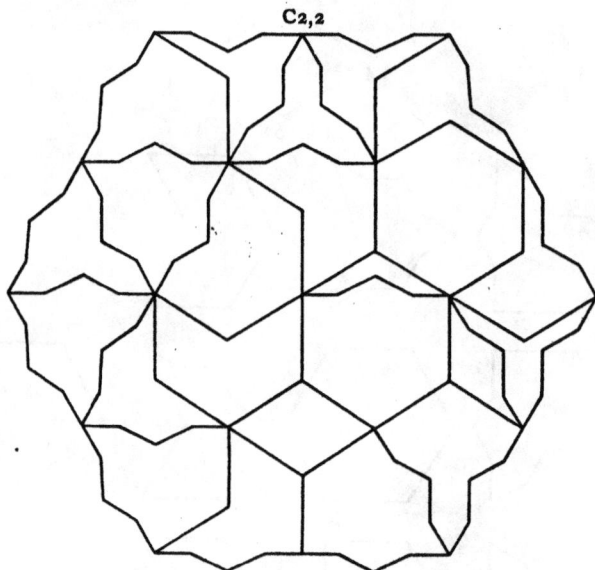

Fig. 65.

Behold, said Pas, a whole dicker of wit.

PEMBROKE, *Arcadia*, p. 393.

42. For the transformation of the 10-piece set on the contact system $C1,1,1$ we may take the colours as in fig. 66 *a*

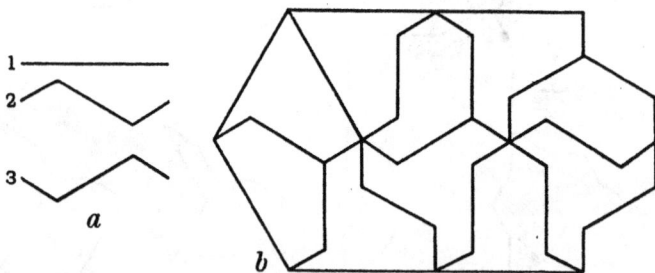

Fig. 66.

and we find for the first figured assemblage $C1,1,1$ $B1,1,6$, the fig. 66 *b* shewing the shapes of the transformed pieces and of the boundary for the particular case.

43.   Skurffe by his nine-bones swears, and well he may,
All know a fellon eate the tenth away.

HERRICK.

Similarly for the contact system $C2,1$, taking the colours as in fig. 67 *a* for the first figured assemblage $C2,1$ $B4,4,0$ we have fig. 67 *b* shewing the shapes of the transformed pieces and of the particular boundary. One piece vanishes.

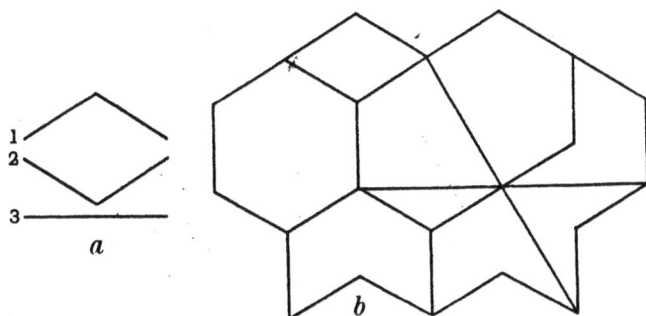

Fig. 67.

44. For the remaining contact system $C1,2$ we may take the colours as in fig. 68 *a*, and for the first figured assemblage $C1,2$ $B5,3,0$ we obtain fig. 68 *b* giving the shapes of the pieces which differ from those in the preceding case. Moreover there are ten of them instead of nine.

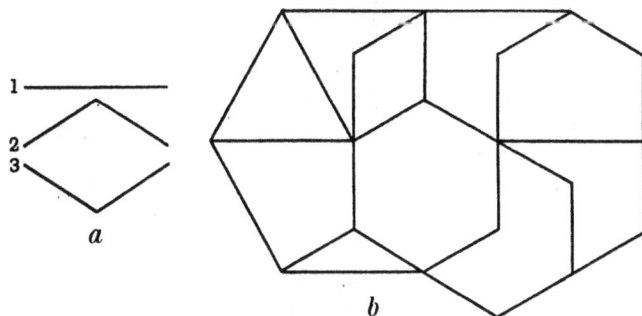

Fig. 68.

45. For the transformation of the 13-piece set we may for the contact system $C1,1,1,1$ select the pieces from those trans-

formed from the complete set of 24 for the same contact system.
The second figured arrangement is then as in fig. 69.

C1,1,1,1 B3,3,3,0

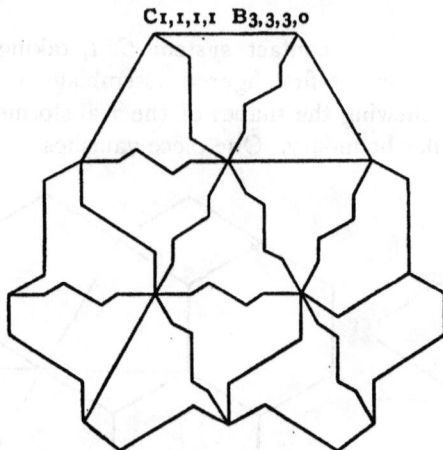

Fig. 69.

46. For the contact system C1,2,1, viz. 1 to 1, 2 to 3, 4 to 4,
we may take the colours as in fig. 70 a; we have then fig. 70 b.

C1,2,1 B1,4,4,0

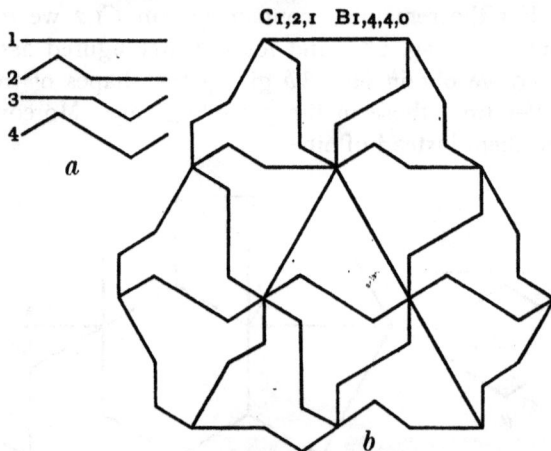

a

b

Fig. 70.

Affirm'd the trigons, chopp'd and changed.
                                    Hudib. II. iii.

47. Coming next to the five-colour triangle we choose, for
the case C1,1,1,1,1, the colours as in fig. 71. The twenty pieces

are as in fig. 72 and the assemblage for $C1,1,1,1,1$ $B12,0,0,0,0$ is as in fig. 73.

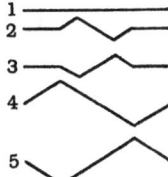

Fig. 71.

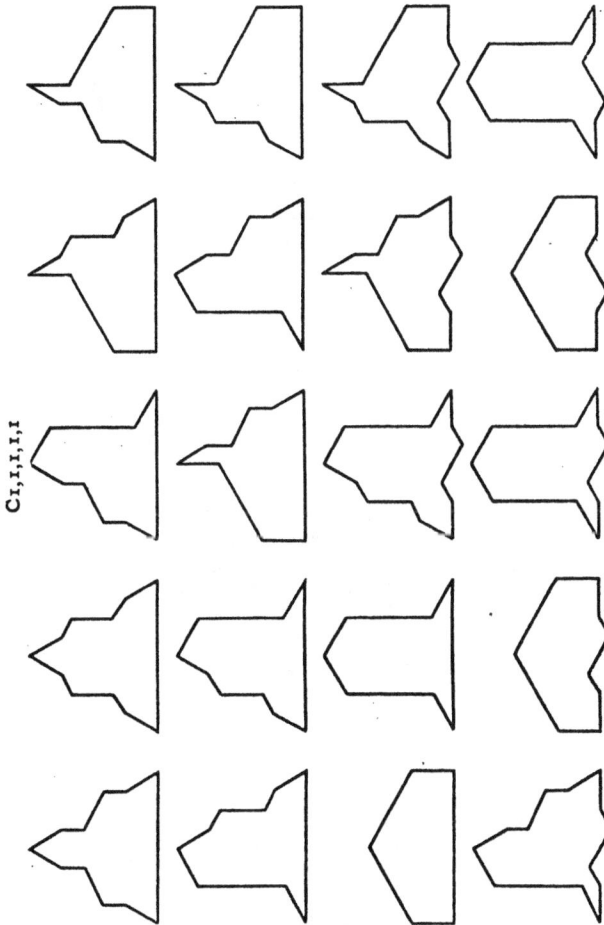

C1,1,1,1 B12,0,0,0

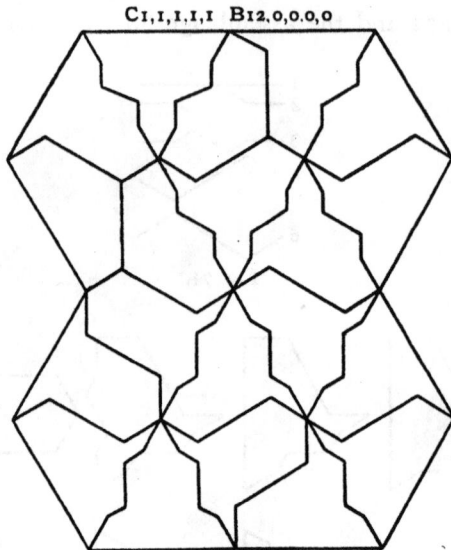

Fig. 73.

48. For the system $C1,1,1,2$, taking the colours as in fig. 75 a, the pieces are as in fig. 75 b, the assemblage as in fig. 74.

C1,1,1,2 B12,0,0,0

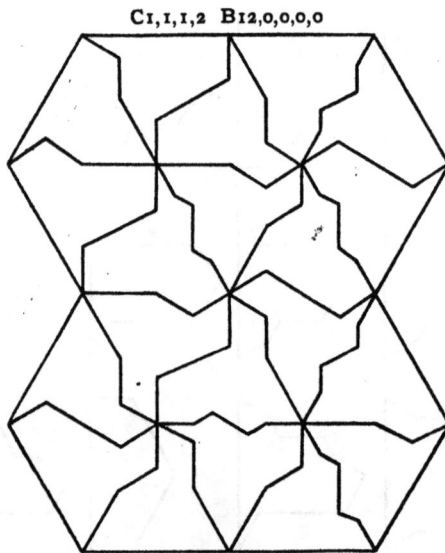

Fig. 74.

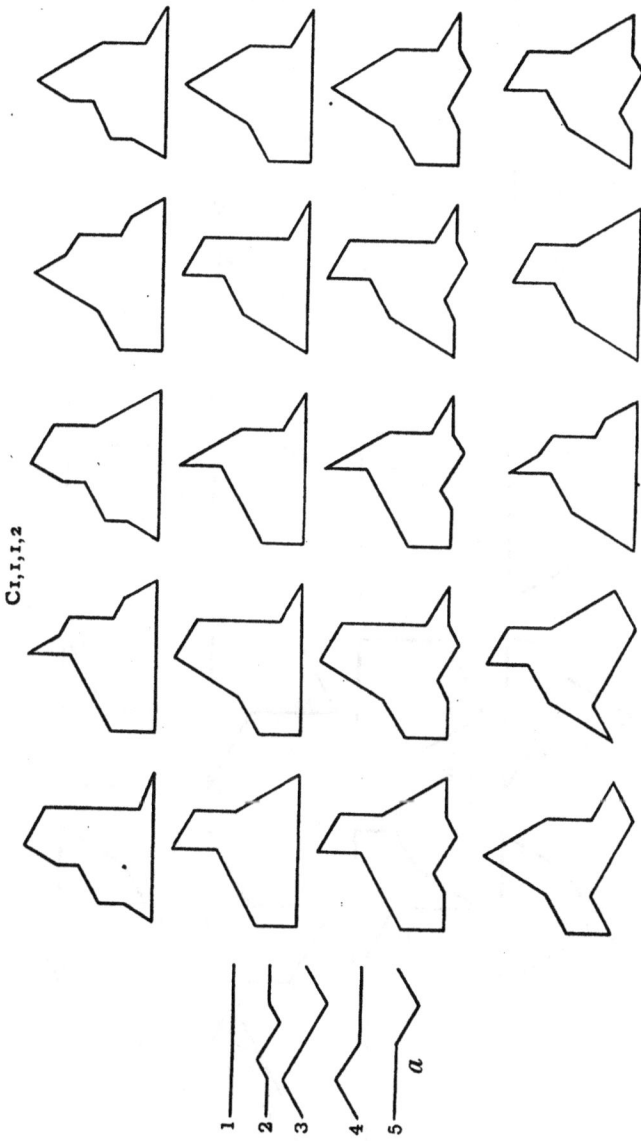

Fig. 75 *b*.

49. For the system $C1,2,2$, taking the colours as in fig. $76\,a$, the pieces are as in fig. 76, and the assemblage for $C1,2,2$ $B12,0,0,0,0$, as in fig. 77.

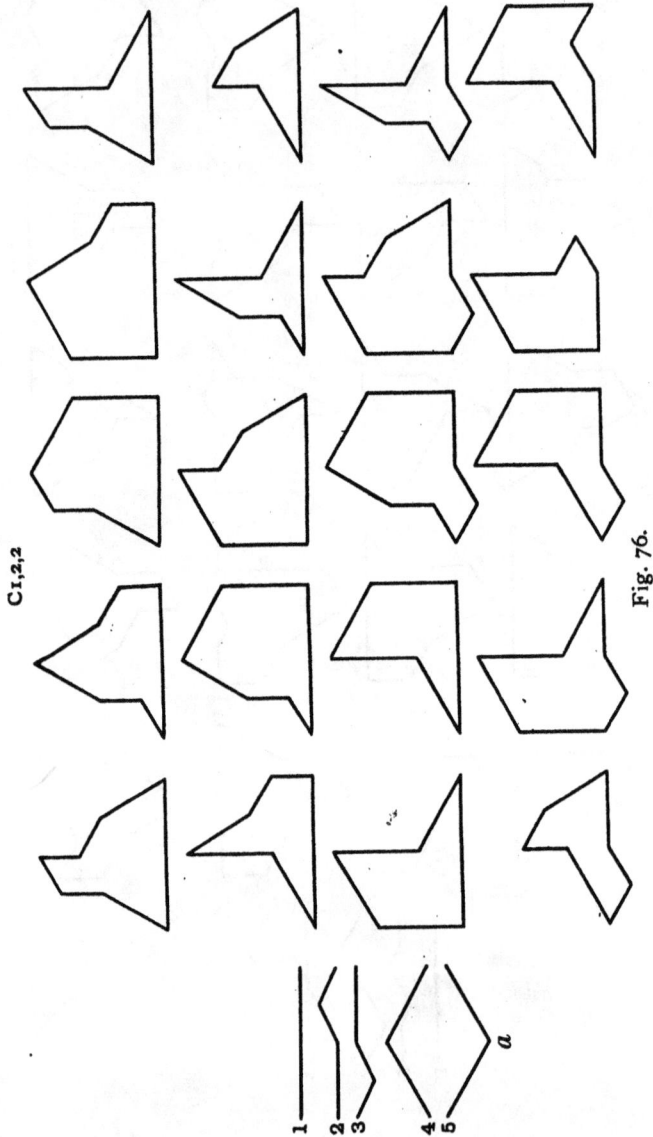

$C1,2,2$

Fig. 76.

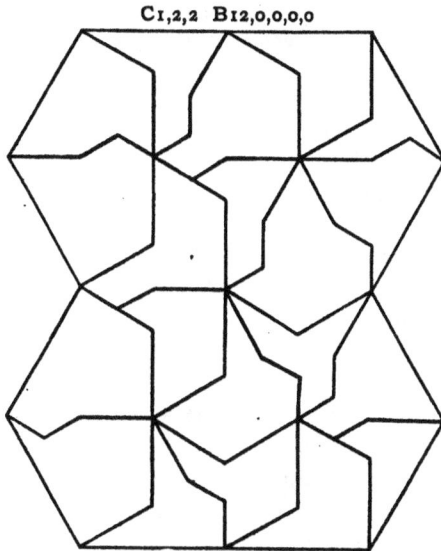

Fig. 77.

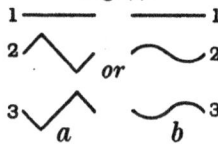

Fig. 78.

(See page 68.)

Has matter more than motion? Has it thought,
Judgment and genius? is it deeply learn'd
In mathematics?

The Consolation.

50. In the case of pieces based upon the square the compartment boundaries must be modified so as to bear a convenient relation to the angles of the parallelogram of the compartments. The parallelogram is here a square and an angle of 45° should be often in evidence. For the contact system $C_{1,1,1}$ suitable boundaries for the 24-piece set may be as in fig. 78 *a* or *b*, p. 67.

The system of pieces for the first of these is given in fig. 78.

For any given boundary defined by colours as in Part I, the figure into which the pieces may be assembled can be drawn.

Thus, as an example, if we take a type and variety such that the colours 2, 3 occur alternately we find the figure 79 *b* and so on. In each case, we make the boundary transformation.

$C_{1,1,1}$ $B_{10,10,0}$

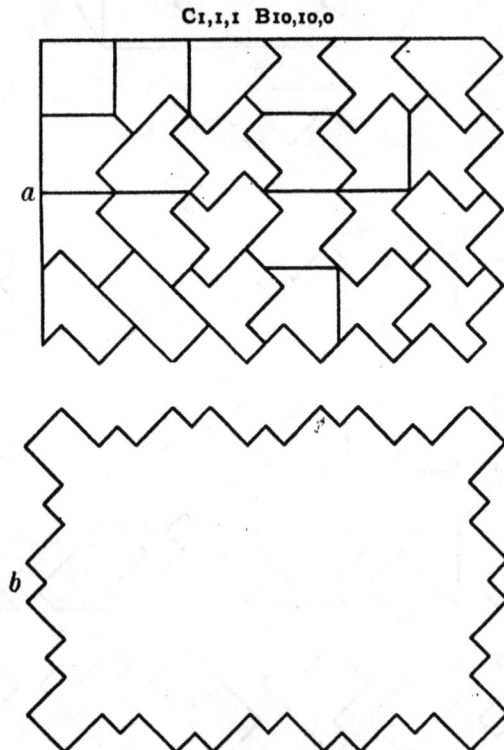

Fig. 79.

A great many symmetrical figures can usually be drawn, with the exercise of a little ingenuity, for any contact system.

The assemblage for $C1,1,1$ $B10,10,0$ is given in fig. 79 a.

After mutch counsayle and great tyme contrived in their several examinations.

Pal. of Pleas. D. d. 2.

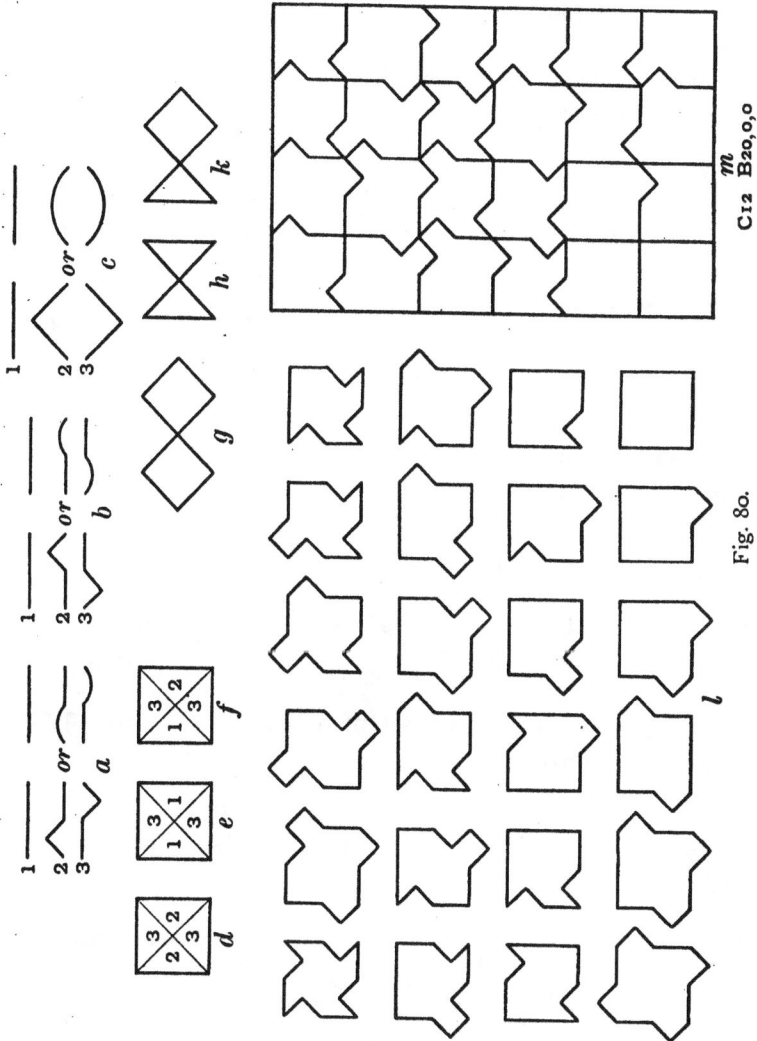

Fig. 80.

The power and corrigible authority of this lies in our wills.
                                            Othello, I. iii.

51. With the contact system $C_{1,2}$, viz. 1 to 1, 2 to 3, we have a good choice for the compartment boundaries; we may take, for example, the colours as in fig. 80 a, b, c.

In the last straight line system of boundaries the piece, which has every compartment coloured 3, vanishes so that the set is one of 23 pieces.

A reference to the boundary types of Part I shews that the set will fit into a large number of symmetrical boundaries.

It should be noted that this 23-piece set is really not suitable because the pieces in fig. 80 d, e, f transform into those in fig. 80 g, h, k, each of which, consisting of two pieces meeting at a point, cannot be handled when constructed in cardboard or wood, but only in diagrams.

These pieces must always be avoided.

The twenty-four pieces are as in fig. 80 l and for the assemblage $C_{1,2} B_{20,0,0}$ we have fig. 80 m.

52. In the case of the 20-piece set of Pastime no. 20 and the contact system $C_{1,1,1}$ we have only to discard the pieces in fig. 81 and then to transform the coloured boundaries in the

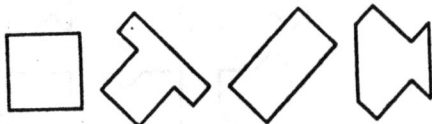

Fig. 81.

usual manner to obtain a set of figures into which the pieces may be assembled.

53. For the contact system $C_{1,2}$, viz. 1 to 1, 2 to 3, we discard from the corresponding set of 24 pieces those shewn in fig. 82 and proceed in the same way.

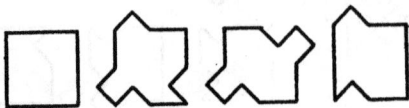

Fig. 82.

54. For the contact system $C2,1$, viz. 1 to 2, 3 to 3, we may discard from the set of 24 pieces corresponding to $C1,2$ the four shewn in fig. 83 and take the correspondence between colour and boundary to be that shewn in the same figure.

Fig. 83.

The reader will find the 20-piece set very interesting but the pieces appear to be difficult to assemble, compared with some of the other sets.

Nihil tam difficile est quin quaerendo investigari possiet.

TERENCE.

55. The 16-piece set transformed for the contact system $C1,1,1$ is shewn in fig. 84 made up into a square of boundary type $B16,0,0$.

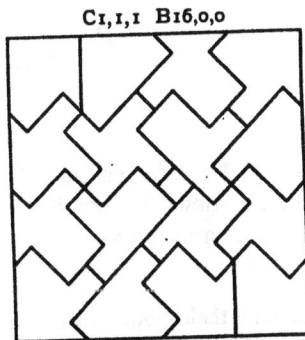

C1,1,1 B16,0,0

Fig. 84.

These same pieces may be fitted into a number of symmetrical boundary lines as will be evident to the reader on comparison with the colour schemes of Part I.

56. The reader will probably have no difficulty in dealing with the contact system $C2,1$, viz. 1 to 2, 3 to 3. The author has not particularly examined it, but he recommends it with confidence.

Sempre avviene
Che dove men si sa, più si sospetta.

MACHIAVELLI.

57. He has however put the contact system $C1,2$, viz. I to I, 2 to 3, through much experimental work, and will deal with it in some detail.

He transforms through the correspondence given in fig. 85 *a*.

It is shewn in the form of 15 pieces since the 16th piece derived from fig. 85 *b* vanishes, but it still must be regarded as a 16-piece set forming up into a square 4 × 4.

The assemblage for $C1,2$ $B16,0,0$ is given in fig. 85 *c*.

$$C1,2 \quad B16,0,0$$

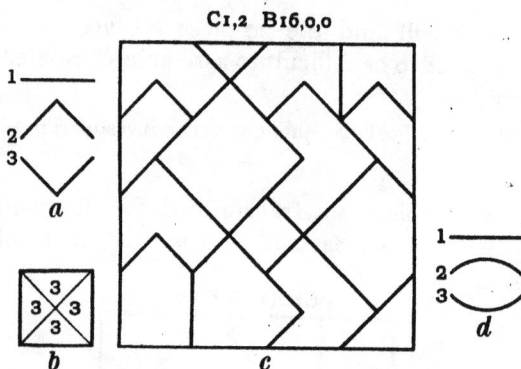

Fig. 85.

Fifteen symmetrical boundaries, out of a large number that may be constructed, are shewn in figs. 86, 87 and 88, inside each of which the same pieces may be assembled.

> Much design
> Is seen in all their motions, all their modes:
> Design implies intelligence and art.
>
> *The Consolation.*

> And soo they thre departed thens and rode forth as faste as ever they my3t tyl that they cam to the forbond of that mount.
>
> *Morte d'Arthur*, I. 139.

For the transformation which has the correspondence of fig. 85 *d* the pieces are shewn assembled in a square and three diagrams of boundaries are also given in fig. 89.

> Mark how the labyrinthian turns they take,
> The circles intricate and mystic maze.
>
> *The Consolation.*

Fig. 86.

Fig. 87.

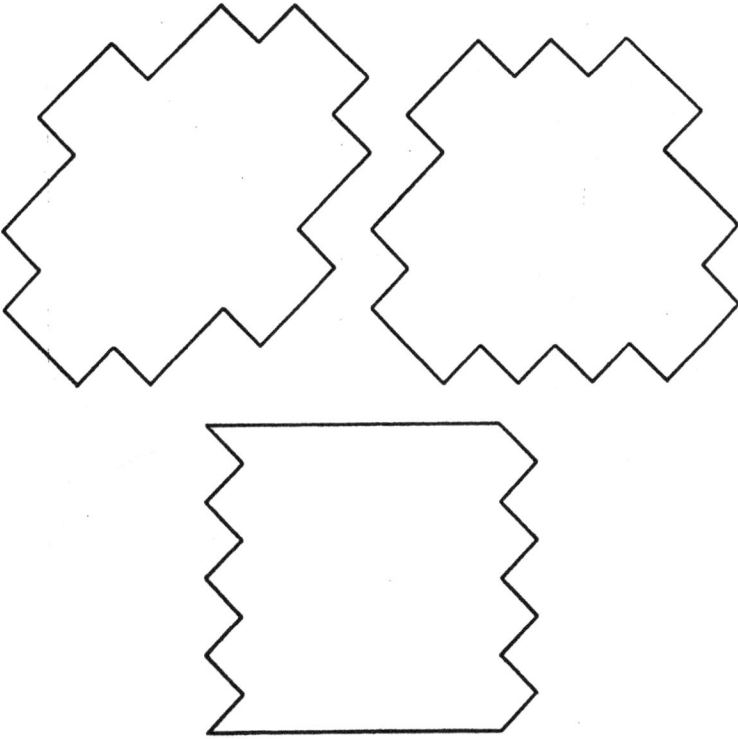

Fig. 88.

The 15-piece set for the contact system $C1,1,1$ is shewn transformed and put up into the rectangle of fig. 90 and other boundaries for the same pieces are shewn in fig. 91.

For the contact system $C1,2$ we have similar results as in fig. 92.

For the 9-piece set we may, for the system $C1,1,1$, transpose to the set in the two upper rows of fig. 93 and for the system $C1,2$ to the set in the two lower rows of the same figure.

For the system $C1,1,1$ a symmetrical boundary $B8,2,2$ is shewn in fig. 94.

For the system $C1,2$ symmetrical boundaries are shewn in fig. 95.

Fig. 89.

Fig. 90.

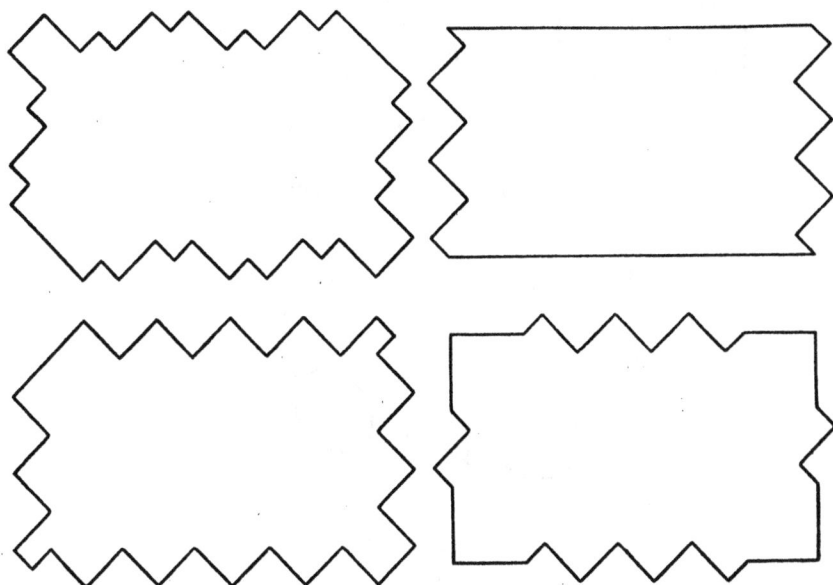

Fig. 91.

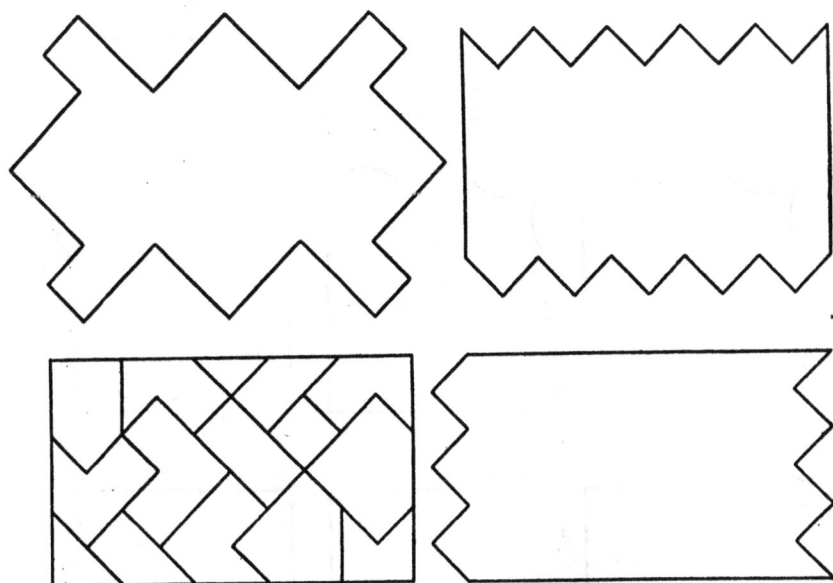

C1,2 B16,0,0   Fig. 92.

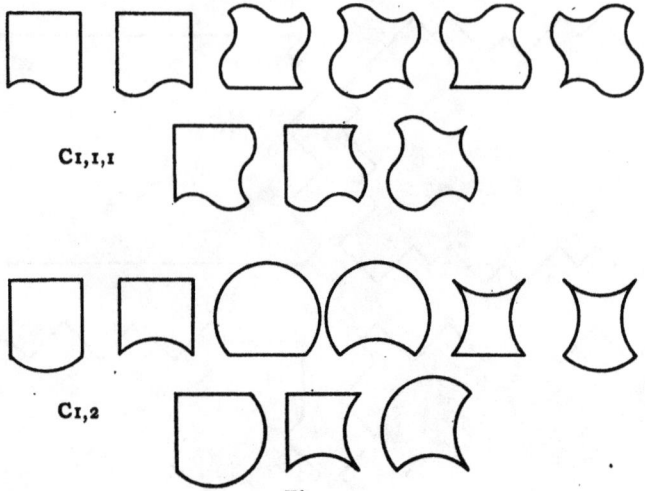

C1,1,1

C1,2

Fig. 93.

Fig. 94.
C1,1,1 B8,2,2

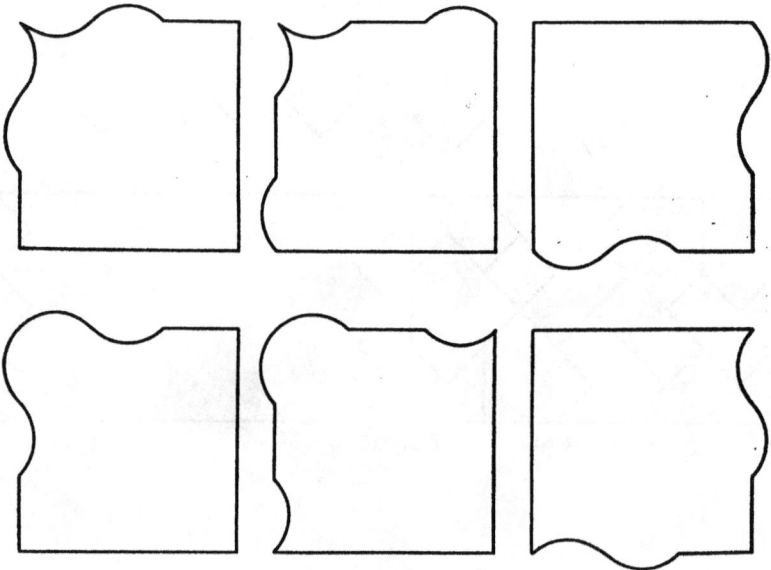

Fig. 95.

58. In fig. 96 *a* the set of 24 squares involving five colours unrepeated, are shewn assembled for the contact system 1 to 1, 2 to 3, 4 to 5 and a transformation according to fig. 96 *b* is shewn in fig. 96 *c*. The symmetry about the central horizontal line is to be remarked.

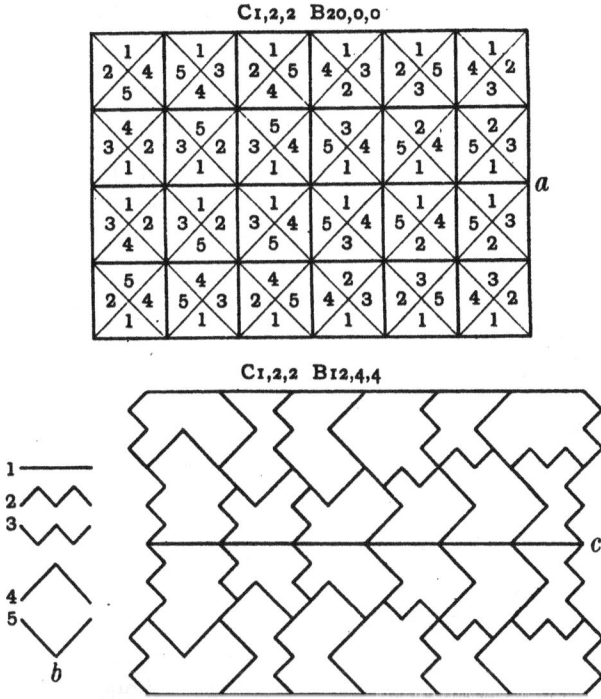

C1,2,2 B20,0,0

1 2×4 5	1 5×3 4	1 2×5 4	1 4×3 2	1 2×5 3	1 4×2 3
4 3×2 1	5 3×2 1	5 3×4 1	3 5×4 1	2 5×4 1	2 5×3 1
1 3×2 4	1 3×2 5	1 3×4 5	1 5×4 3	1 5×4 2	1 5×3 2
5 2×4 1	4 5×3 1	4 2×5 1	2 4×3 1	3 2×5 1	3 4×2 1

a

C1,2,2 B12,4,4

1 ——
2 /\/\/\
3 \/\/\/

4 /\
5 \/

b

c

Fig. 96.

It is obvious that there is ample room for experiments with this interesting set.

PART III

THE DESIGN OF 'REPEATING PATTERNS' FOR DECORATIVE WORK

The story without an end that angels throng to hear.
                                        M. F. TUPPER.

59. The ideas which have been prominent in Parts I and II lead to a most interesting Pastime—the design of repeating patterns for various kinds of decoration. In Part I the notion was to connect colours with the compartments of various polygons, usually triangles and squares, in such wise as to realise every possible combination of colours on the compartments. The set thus formed has the property that no two pieces are alike. There are no duplicates. Smaller sets are chosen from these according to definite principles and laws but the pieces are in every case differently coloured. In the transformations of Part II the passage is made from pieces of the same shape but differently coloured to pieces of the same colour but differently shaped. The transformations do not affect the cardinal property that no two pieces of any set are alike.

It was shewn that the pieces of a set can be assembled so as to fit inside a boundary which can be specified as to colour or shape and that, for a given set of pieces, the different boundaries that can be predicted exist, for various contact systems, in very large numbers.

We now examine what can be done with pieces which, far from being all different, are all exactly similar in shape and size.

Certain pieces of the same size and shape can, everybody knows, be fitted together so as to completely cover any floor, wall, ceiling or other flat surface, and no attention need be paid to the boundary. The pieces, when assembled, can be cut along any desired boundary. For practical purposes the boundary may be ignored.

The simplest repeating patterns are shewn in fig. 97 and are met with everywhere.

We make a systematic search for pieces of other shapes which possess the same property of completely covering any flat surface by simple repetition.

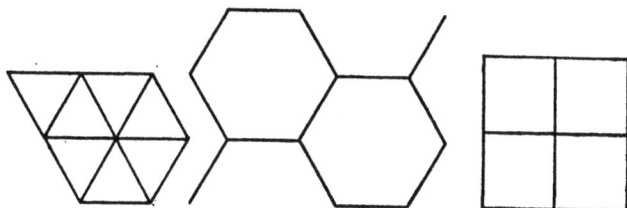

Fig. 97.

These 'repeating patterns' are everywhere visible in considerable variety.

Public buildings and private homes exhibit examples on floors, walls and ceilings. Street pavements, palings, furniture, wearing apparel, woven fabrics and artistic designs of all sorts bring them constantly before our eyes.

We feel that the ideas brought before the reader in this book will enable him to take quite a fresh interest in these matters.

Without making any pretence of exhausting the subject we base our study upon the equilateral triangle, the square, and the regular hexagon. These are probably the most important bases.

I saw it budding in beauty; I felt the magic of its smile.

M. F. TUPPER.

60. It is a very trivial fact that the equilateral triangle is a repeating pattern for any flat surface; but one useful remark may be made, viz. that in such an assemblage, a portion of which is represented above, the triangle has always just two aspects or orientations, because the fitting is invariably so that of two adjacent pieces one is always like the other, only upside down, as may be readily seen.

This is not the case in either the square or hexagonal pavements, which present only one aspect.

The equilateral triangle, which as a repeating pattern has two aspects, gives immediate rise to other repeating patterns upon very simple principles.

If we can in any manner separate it into three or more parts of the same size and shape we obtain again a repeating pattern.

For example we can as in fig. 98 from the centre O draw perpendiculars upon the sides and obtain the equal and similar quadrilaterals of which one is $ODBF$, a repeating pattern which in an assemblage has 3 × 2 or 6 aspects.

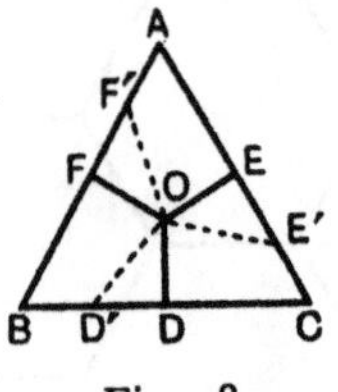

Fig. 98.

Also we may draw the lines OD', OE', OF' making equal angles with OD, OE, OF and obtain the repeating pattern $OD'BF'$.

This system of quadrilaterals has two opposite angles 60° and 120° respectively.

61. Similarly from the square we can at once derive an unlimited number of repeating patterns by drawing lines, straight or not, through the centre as in fig. 99.

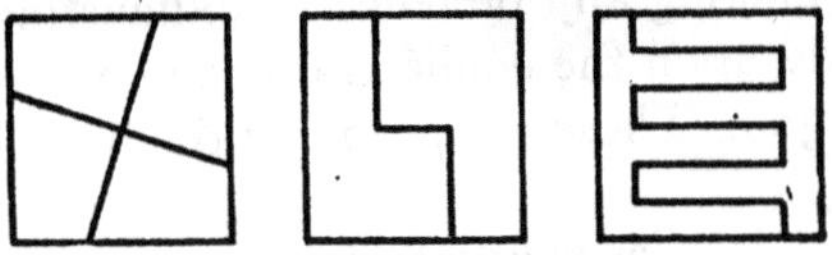

Fig. 99.

The first of these figures shews a separation into two or four parts, the second into two parts, the third into two parts. There is no limit to the number of such separations.

Any number of squares may be placed in contact to form a pattern, and a little consideration shews that it follows that every rectangle is a repeating pattern.

> Who made the spider parallels design,
> Sure as de Moivre, without rule or line?
>
> *Essay on Man.*

If we draw a series of equidistant parallel lines and cut them with *any other* series of equidistant parallels, we see at once that *every parallelogram* is a repeating pattern with one aspect, and since every parallelogram can be separated into two equal and similar triangles; and *vice versa* every triangle is the half of a parallelogram; it follows that *every* triangle is a repeating pattern, with two aspects.

As a general principle, if any repeating pattern can be divided into two or more equal and similar parts, a part thus formed is a repeating pattern *.

62. When we assemble a number of patterns so as to cover a flat space we can always observe combinations of patterns which are themselves patterns. If the pattern has a certain number of aspects in the assemblage and we draw a boundary which includes *one* pattern of each aspect, the figure enclosed by the boundary is a pattern with one aspect.

For example the equilateral triangle has two aspects, and we can draw a regular hexagon which encloses six triangles, three of each aspect. Thence we conclude at once that the regular hexagon is a pattern having one aspect.

63. The regular hexagon may be derived as a repeating pattern from the equilateral triangle in another interesting manner.

We can colour an assemblage of triangles as in fig. 100 *a* with two colours so that each triangle is adjacent to three triangles of a different colour.

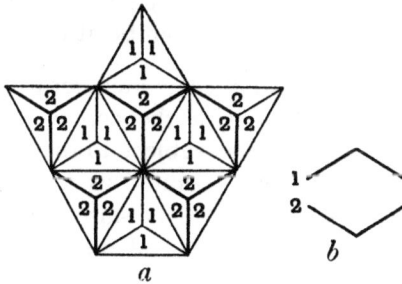

Fig. 100.

It is by others termed a fesse between two *gemels*. And that is as farr from the marke as the other; for a *gemel* ever goeth by paires, or couples, and not to be separated.

R. HOLME, *Academy of Armory*, I. iii. 77.

We can now transform from colour to shape by the method of Part II on the contact system 1 to 2, taking for the colours

* Triangles may be taken to be equal and similar in *plane* geometry when they can be made to exactly coincide by movements in the plane; a movement such as 'turning over' is not in view.

6—2

the boundaries as in fig. 100 *b*, when we find that the triangles coloured 2 entirely disappear and the triangles coloured 1 become regular hexagons.

We have transformed the tessellation, from being triangular to being hexagonal.

64. The mathematical reader will recall that, in space, the cube and the rhombic dodecahedron are similarly associated. In that case we start with a number of cubes and give them two colours as in a three-dimensional chessboard. The cubes coloured 2 are divided into six portions by lines joining the centre to the eight summits; each portion being a pyramid whose base is a face of the cube and whose height is equal to half of the cube's edge. Each cube coloured 1 is adjacent to six cubes coloured 2, and transformation upon the contact system 1 to 2 with pyramidal boundaries causes the cubes coloured 2 to disappear and those coloured 1 to become rhombic dodecahedra.

Space is thus exhibited as partitioned into rhombic dodecahedra, a fact of great importance in crystallography and other parts of applied mathematics and mathematical physics.

The foregoing considerations bring out the importance of designing patterns which possess symmetry about one or more axes; from these other patterns can be at once derived.

65. In Part II we adopted certain forms of compartment boundaries in association with the contact systems, and the design of such boundaries was subject to the transformed pieces being readily handled. This operated as a considerable restriction which can now be removed because our pieces are merely delineated; they are not handled at all. The restriction took two forms.

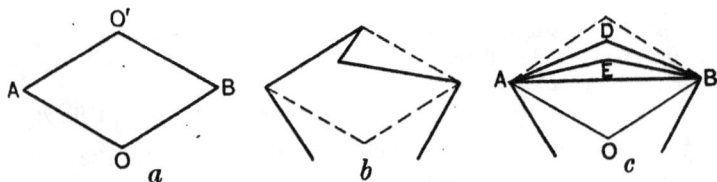

Fig. 101.

A O'BO in fig. 101 *a* being the compartment parallelogram, in drawing the new boundary between *A* and *B* and within the parallelogram we decided that no perpendicular to AB should cut the new boundary in more than one point; because if it did so the piece would be inconvenient to handle. Again it was tacitly agreed that the portion of the piece bounded by the new boundary line and AOB should have no *holes* in it. Both of these restrictions can now be removed and we can deal with shapes such as in fig. 101 *b* and 101 *c* where in the second case there is an inner as well as an outer boundary, the triangular piece AEB being cut out.

THE TRIANGLE BASE

66. Consider the triangle to be divided into compartments numbered 1, 2, 3 so that, travelling clockwise round the figure, the numbers are in ascending order of magnitude as in fig. 102 *a*.

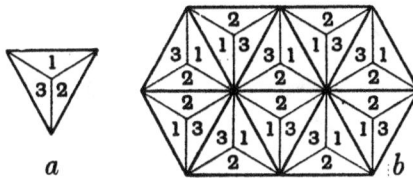

Fig. 102.

It is not quite a trivial remark that any number of triangles identical with this one can be assembled so that the compartments numbered 1, 2, 3 are adjacent to others numbered 1, 2, 3 respectively.

The annexed diagram, fig. 102 *b*, establishes this and shews that the number of aspects of the triangle is not increased by the numbering. We knew before that it could not be less than two. We know now that it remains precisely two.

Also the internal structure is always the same.

The reader who has become acquainted with Part II will observe that the straight line boundaries of the compartments of the triangles can be altered in many ways so that the pieces remain of the same shape and size. In other words the numbered triangular piece can have its shape altered to an indefinite extent and still preserve the property of being a repeating pattern.

The assemblage under examination has the contact system 1 to 1, 2 to 2, 3 to 3, and we may employ any of the boundaries which belong to the system. The numbers 1, 2, 3 need not differ from one another.

The particular cases 1, 1, 1; 1, 1, 2 are included in the discussion.

67. We are about to study the design of repeating patterns and the reader will readily realise that a pretty pattern must always be an object, and a principal object, in the designer's mind. This can best be accomplished by paying attention to symmetry of shape.

We must learn how to select boundaries which will produce symmetrical patterns. This matter must be studied in the case of each contact system and each base.

To assist us we must have a typical boundary before us, say A in fig. 103 a, and associate with it another boundary which we will call the inverse of A and denote by iA.

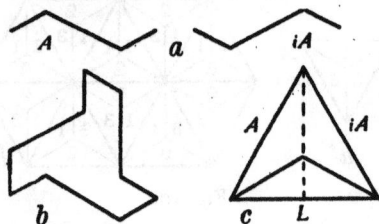

Fig. 103.

iA is the reflexion of A in a mirror placed to its right, *or* it is A rotated about its right-hand extremity, in a plane perpendicular to the plane of the paper, until it occupies the position of iA.

It will be noticed that if A be the original single straight line, inversion does not alter it.

In the first place it is clear that if the numbers 1, 2, 3 are associated with the same boundary A, however we choose A, the piece will possess trilateral symmetry.

For example, for the typical boundary we get fig. 103 b.

The only other kind of symmetry is that about an axis which bisects one of the angles of the triangle; say about the line chain-dotted in fig. 103 c.

As regards the sides which meet at the angle fixed upon, the only possibility is to associate them with the boundaries A, iA respectively.

The reader is reminded that a compartment-boundary is to be viewed from the centre of the base, looking outwards.

In regard to the third side it may remain associated with the single straight line.

We call this *L* for short.

Notice also that the shape is altered when we interchange *A* and *iA*.

For the typical boundary we thus get the two patterns of fig. 104 *a* and *b*.

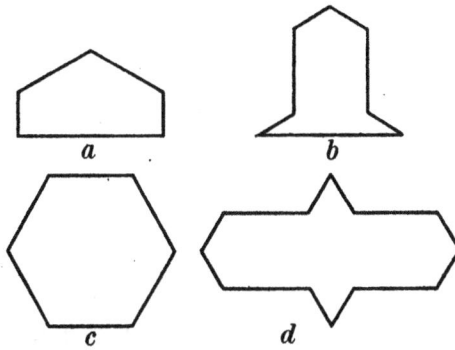

Fig. 104.

Inspection of the numbered diagrams shews that we can pass from them to repeating patterns, fig. 104 *c* and *d*, of which the first is the regular hexagon again. In both cases the sides *L* have been made adjacent. Other patterns are obtained by making the sides *A* or the sides *iA* adjacent in each case.

68. The particular case 2 equal to 3 may be noticed. It is interesting from the circumstance that all pieces of this type may be assembled so as to have either two or six aspects. This is made clear by the diagrams in fig. 105 in which the diagram on the left exhibits two aspects and that on the right six.

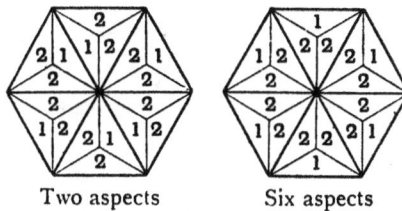

Two aspects   Six aspects

Fig. 105.

In fig. 106 some patterns and assemblages belonging to this system are given and in particular the 'Helmet' pattern is

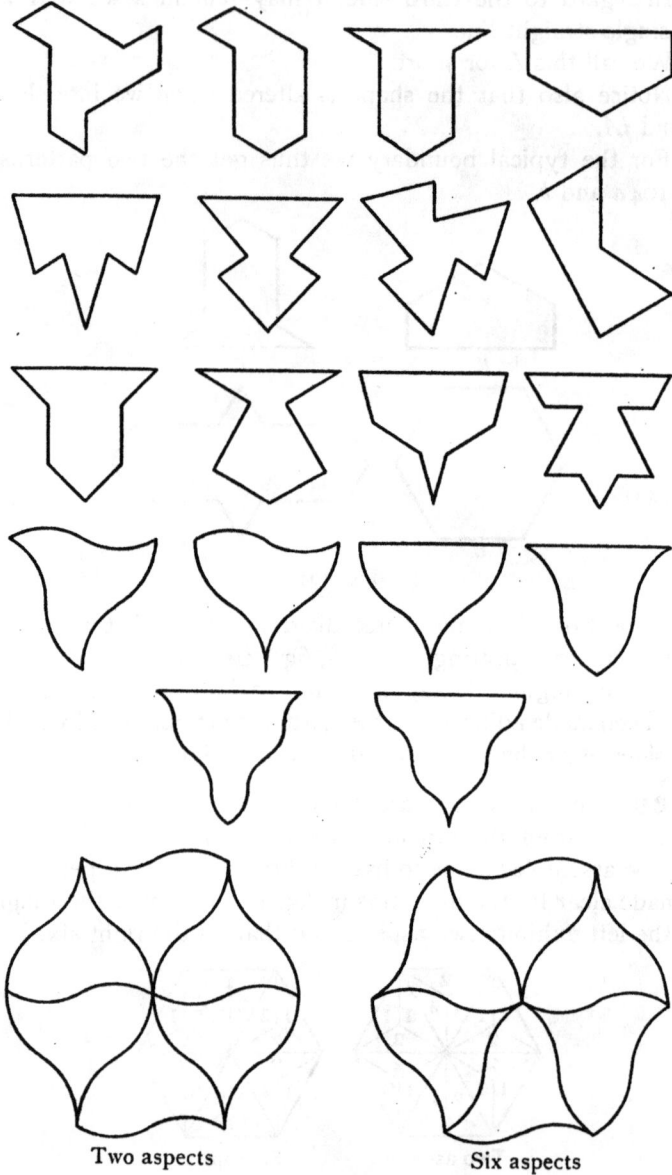

Two aspects   Six aspects

Fig. 106.

shewn assembled so as to exhibit two aspects, on the left, and six aspects, on the right.

Besides the equilateral triangle, since every triangle is a repeating pattern, we may take as base the isosceles triangle in general as in fig. 107 and the isosceles right-angled triangle in particular.

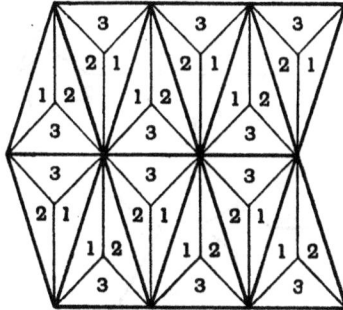

Fig. 107.

The diagram shews that, for this contact system, the treatment is the same. The reader will find no difficulty with this general case.

> So that the art and practic part of life
> Must be the mistress to this theoric.
> > *Hen. V.* i. i.

69. A second and entirely different system of patterns is obtainable from the contact system 1 to 1, 2 to 3 which we call C1,2.

Inspection of the depicted assemblages of six triangles in the form of a hexagon as shewn in fig. 108 shews that the piece has six aspects in each of two constructions, which are at the disposition of the designer.

Part II again shews us how to alter the shape of the piece so that it will remain a repeating pattern.

> *Herm.* Methinks I see these things with parted eye,
> When ev'rything seems double.
> *Hel.* So methinks:
> And I have found Demetrius like a *gemel*,
> Mine own, and not mine own.
> > *Mids. N. Dr.* iv. i.

The boundary of compartment 1 is altered in any manner which appertains to the first system already dealt with, whilst

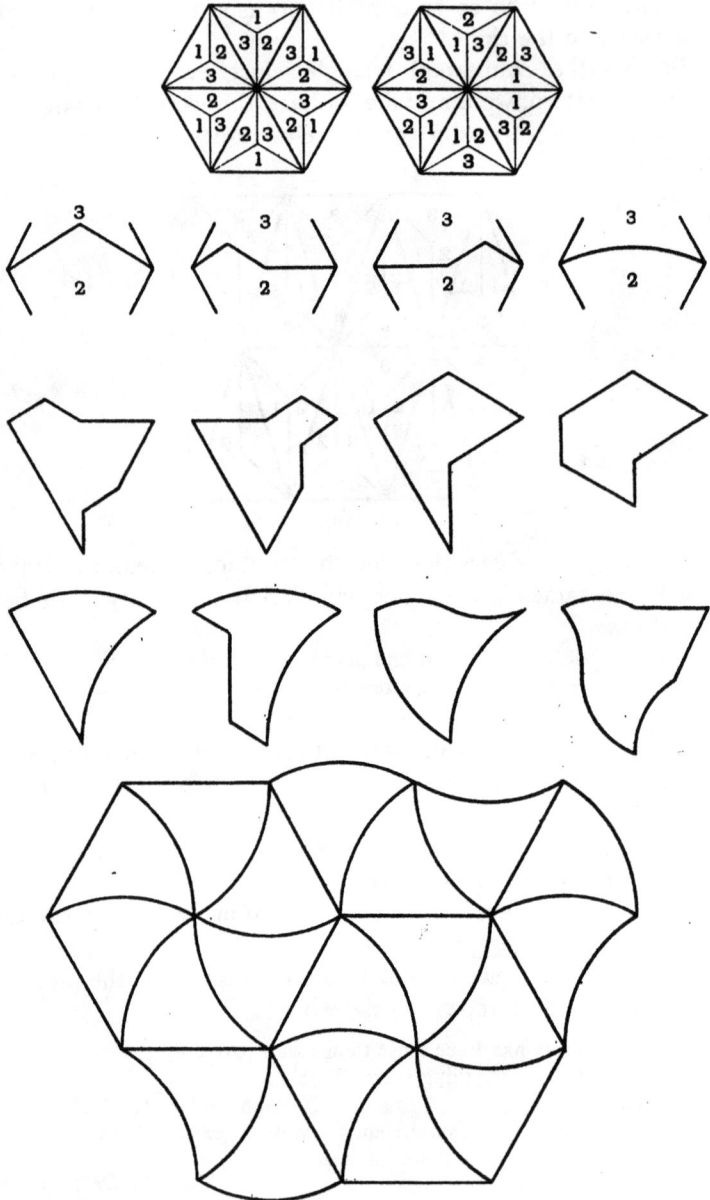

Six aspects

Fig. 108.

for compartments 2 and 3 we have the boundaries of Part II as in the numbered row of fig. 108 where the 3rd is the inverse of the 2nd, while the 1st and 4th are self-inverse.

Some examples of repeating patterns, issuing from this system, and an assemblage are given in the same fig. (108).

Symmetry can be secured by taking six identical pieces as in the annexed diagram, fig. 109 *a*, which is a well-known pattern composed of three hexagons.

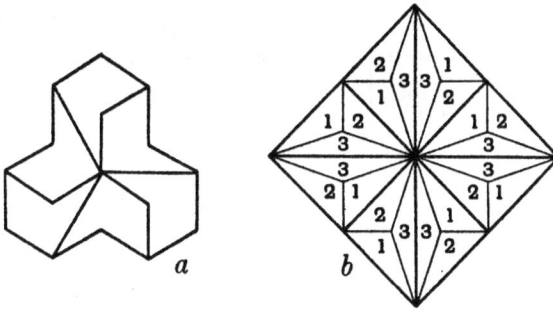

Fig. 109.

The isosceles right-angled triangle is also available for this contact system, as shewn by the diagram of fig. 109 *b*. The system is 1 to 2, 3 to 3, and the derived patterns have four aspects when assembled.

THE SQUARE BASE

For the wals glistered with red marble and pargeting of divers colours, yea all the house was paved with checker and tesseled worke.

KNOLLES' *Hist. of Turks.*

But I of these will wrest an alphabet,
And by still practice learn to know thy meaning.

Tit. Andron. III. ii.

70. We recall that the piece does not, of necessity, have more than one aspect.

Numbering the compartments as in fig. 110 *a* we can always assemble according to the contact system 1 to 1, 2 to 2, 3 to 3, 4 to 4 which is denoted by $C_{1,1,1,1}$.

This is shewn in fig. 110 *b*, a diagram which can be repeated indefinitely.

The arrangement proves that it is possible in one way only and that the piece has two aspects, one being the other rotated through two right angles.

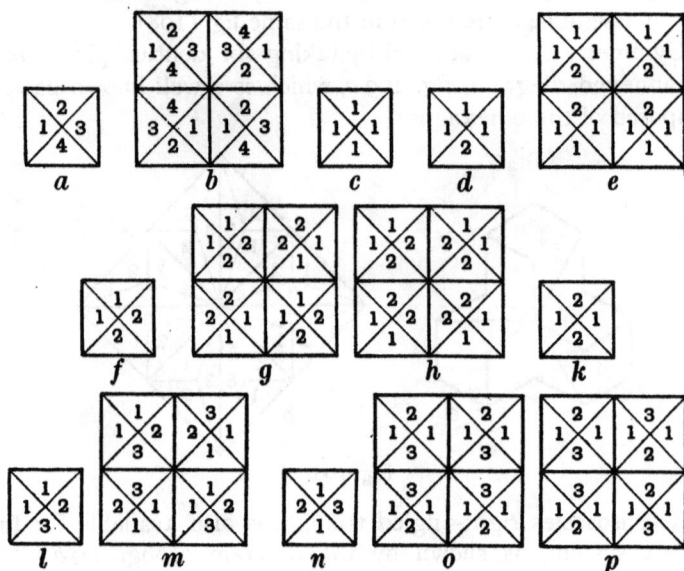

Fig. 110.

In the particular cases: fig. 110*c* has one arrangement with one aspect of piece.

Fig. 110*d* has only one symmetrical arrangement which deserves consideration as shewn in fig. 110*e* with two aspects of piece.

The piece in fig. 110*f* has two arrangements shewn in fig. 110*g* (two aspects), in fig. 110*h* (four aspects).

The piece in fig. 110*k* has obviously one arrangement with one aspect of piece.

The piece in fig. 110*l* has one arrangement with two aspects as shewn in fig. 110*m*; and finally the piece in fig. 110*n* has the arrangements of fig. 110, *o* and *p*, each with two aspects.

71. To construct repeating patterns we take four shapes of boundary drawn from the first system of boundaries, taking care to choose those that are appropriate to the square base.

We are restricted to the square $AOBO'$ in fig. 111 and no circular arc employed must lie outside of it at any portion of its course between A and B. When the arc passes through A the centre from which it is struck should be taken upon either AO'

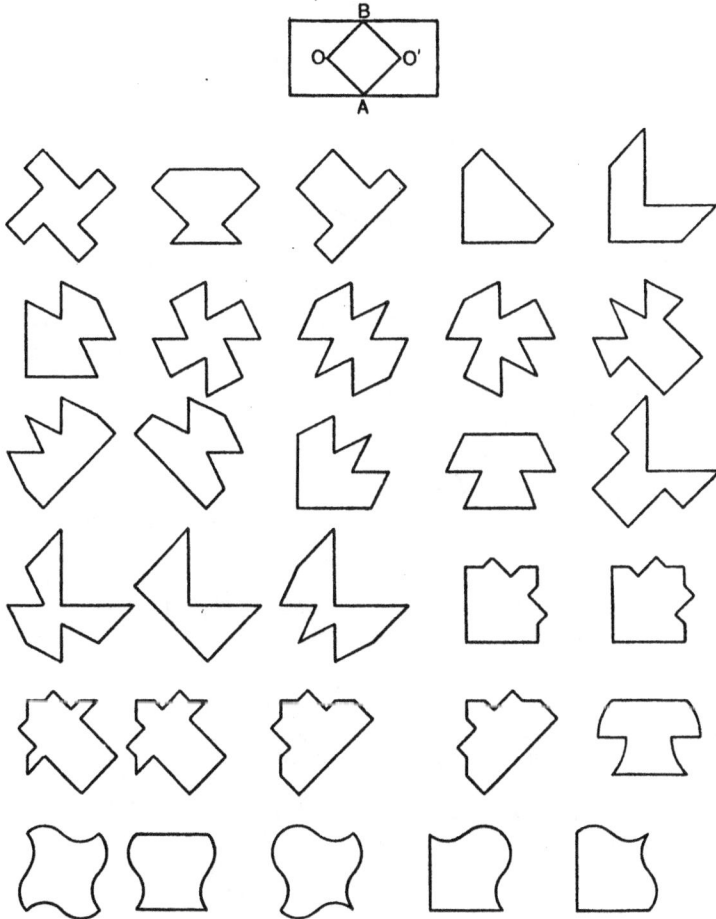

Fig. 111.

or AO produced if necessary so that it touches OA or $O'A$ at the point A. Part II points out convenient forms and leads us to a variety of patterns of which samples are shewn in fig. 111 and assemblages in fig. 112 where a and b exhibit two aspects and c four.

Seizes the prompt occasion, makes the thought
Start into instant action, and at once
Plans and performs, resolves and executes!

HANNAH MORE.

Fig. 112.

Two other assemblages are shewn in fig. 113 where fig. *a*
exhibits two aspects and *b* four.

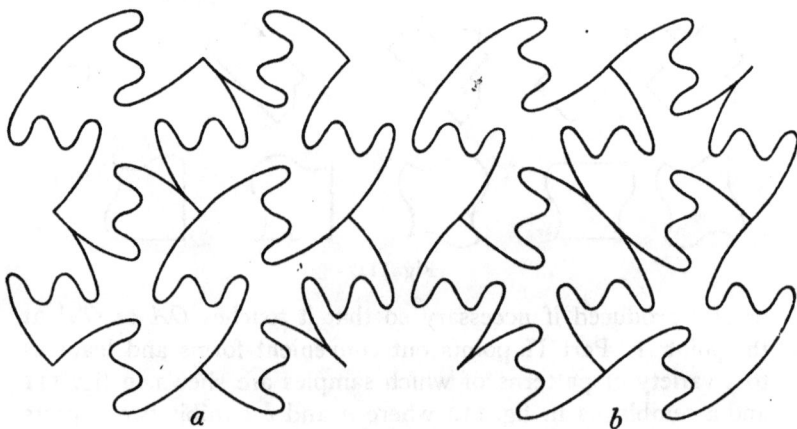

Fig. 113.

SYMMETRY OF PATTERN

Veluti in speculum.

Let no face be kept in mind,
But the fair of Rosalind.

As You Like It, III. ii.

72. We now consider some more principles which lead to symmetry of pattern.

We take four compartment boundaries as in fig. 114 *a*; we have already defined *iA*; we call the one beneath *iA* the complement of *A* and denote it by *cA*. It is called the 'complement' because the compartment with the boundary *cA* will fit into (lie adjacent to) the compartment with the boundary *A*.

So also the boundary beneath *A* is called the complement of the inverse of *A*.

It will be noticed that *ciA* and *cA* are reflexions of *A* and *iA* in the horizontal chain-line delineated.

Also that *icA* is the same as *ciA* and that *ccA*, *iiA* both leave *A* unaltered.

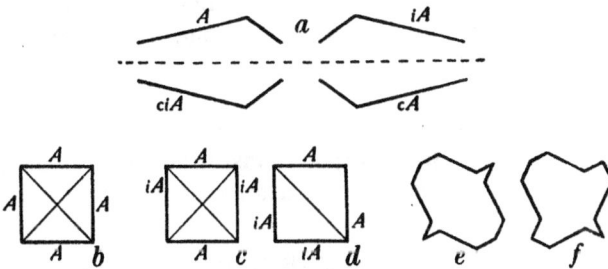

Fig. 114.

cA and *A* on the one hand and *ciA* and *iA* on the other have to do with the second kind of contact between compartments.

We are immediately concerned only with the first kind of contact and we see that we may have, as in fig. 114*b*, the same boundary to each compartment giving quadrilateral symmetry.

Or we may have the designs with diagonal symmetry in fig. 114*c* and *d* where the first has symmetry about both diagonals, because both bisect angles contained by *A* and *iA*; and the second has only symmetry about the diagonal drawn because the other does not satisfy the *A* and *iA* condition.

Examples are shewn in fig. 114*e* and *f*.

We will have, if this fadge not, an antick. I beseech you, follow.
Love's L. L. v. i.

73. To obtain symmetry about a straight line through the centre parallel to a side the form that must be taken is that in fig. 115 *a* of which an example is fig. 115 *b*.

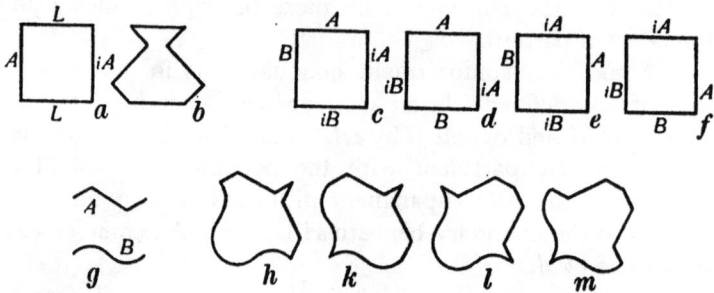

Fig. 115.

With two boundary forms *A, B* we have four forms, fig. 115 *c, d, e, f*, each of which has symmetry about one diagonal.
Examples for the boundaries in fig. 115 *g* are shewn in fig. 115 *h, k, l,* and *m*.

74. The next contact system to study is that in which we have 1 to 1, 3 to 3, 2 to 4 and inspection of the diagram in fig. 116 shews that there is one arrangement with two aspects of piece.

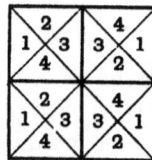

Fig. 116.

This is also the case when 1 and 2 are identical.
For the boundaries of compartments 2 and 4 we choose from those brought forward in Part II that are appropriate to the square base.
Patterns of quite a new character emerge, as is evident from the samples in fig. 117.

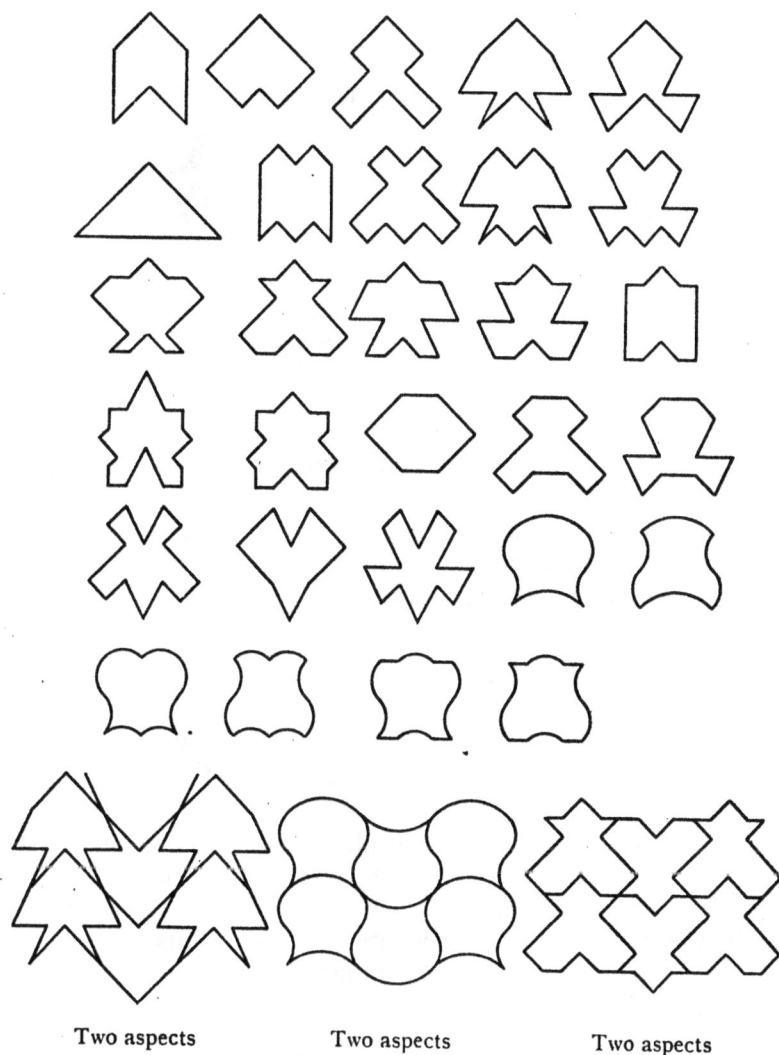

Two aspects   Two aspects   Two aspects

Fig. 117.

75. Here the only possible symmetry is that about an axis through the centre parallel to a side. The general forms are as in fig. 118 *a* and *b*, where *B*, *A* belong to the first and second kinds of contact respectively.

In particular either *A* or *B* may have the form *L*.

A must also be symmetrical about its extremities, so that $A = iA$.

As an example, for the forms in fig. 118 c we find fig. 118 d, a piece with windows.

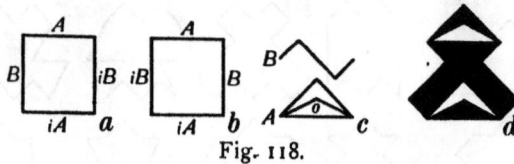

Fig. 118.

(Deh come è ver che) subito travato
Il bello piace a chi non è malato.

BRACCIOLINI.

76. A few minutes trial will convince the reader that the contact system 1 to 1, 2 to 2, 3 to 4 is impossible unless 1 and 2 are identical.

For the case specified we have the arrangement shewn in the diagram on the left of fig. 119 involving four aspects, and no other essentially different from it.

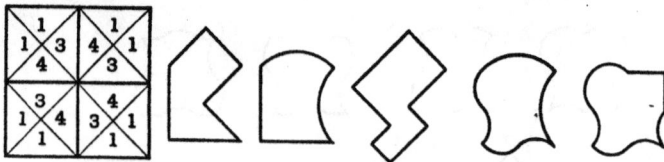

Fig. 119.

This system does not lend itself to interesting symmetrical patterns, for two reasons. In the first place 3 and 4 are not opposite compartments, and in the second the inverse principle cannot be applied to the two identical compartments 1.

However, a few forms are given in fig. 119.

77. A more interesting set of patterns arises from the contact system 1 to 3, 2 to 4. Here the contacts are of oppositely situated compartments and the inverse principle is also available.

The arrangement is shewn in fig. 120 a. It involves only one aspect of piece.

It can be repeated to any extent on all sides.

Exceptionally the piece in fig. 120 b with the contact system 1 to 3, can be assembled in three different ways having one, two and four aspects respectively.

This is shewn by the diagrams 120 *c* with one aspect, 120 *d* with two aspects and 120 *e* with four aspects.

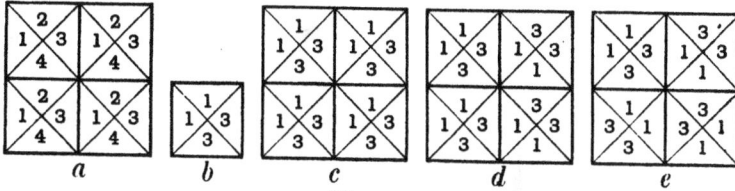

Fig. 120.

Some patterns and assemblages are shewn in fig. 121 where *a* has one aspect, *b* two aspects, *c* four aspects and *d* one aspect.

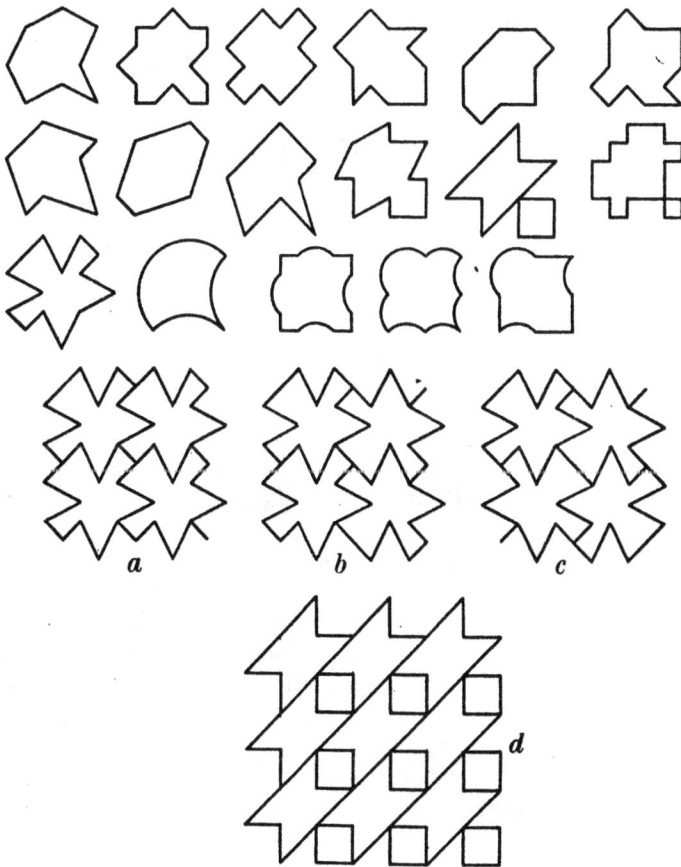

Fig. 121.

78. The three symmetrical forms are shewn in figs. 122 *a*, *b*, and *c* where for the first of these *A* must be identical with its inverse.

Examples of the last two are figs. 122 *d* and *e*, and of the first fig. 122 *f*.

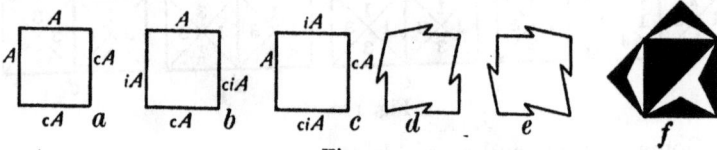

Fig. 122.

79. Lastly we have the contact system 1 to 4, 2 to 3. The arrangement is as in fig. 123.

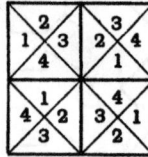

Four aspects

Fig. 123.

The arrangement can be, clearly, repeated to any extent.

Some examples of the patterns that arise are given in fig. 124 where assemblage *a* has two aspects and assemblage *b* has four.

The reader will observe that the tessellation (fig. 124 *b*) is composed of two sets of hexagons at right angles to one another.

Each hexagon contains four equilateral pentagons, one of each aspect.

80. For symmetry about an axis parallel to a side we have fig. 125 *a* where *A* is a form which is identical with its inverse; and for symmetry about a diagonal the two forms in fig. 125 *b* and *c* when *A* is as in fig. 125 *d*, the triangle numbered zero being cut out; the pattern arising from the last form but one is curious, as shewn in fig. 125 *e*.

It has four aspects and requires four colours when assembled.

In the assemblage shewn in fig. 125 *f* the numbers denote different colours.

Pieces having the same aspect are given the same colour.

Fig. 124.

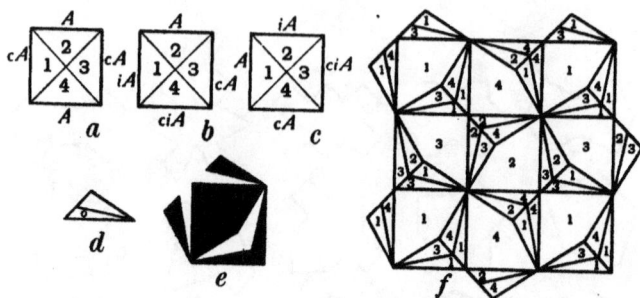

Fig. 125.

Parfois dans un coin triste et noir pousse une fleur.

FRANÇOIS COPPÉE.

See'st thou the gaze-hound! how with glance severe
From the close herd he marks the destin'd deer.

STEELE'S *Miscellanie.*

81. The tessellation of pentagons that has been depicted is one of the most remarkable that can be met with in the subject.

If the patterns that have been given above, for the contact system under examination 1 to 4, 2 to 3, be inspected they will be found to include three convex pentagons, out of an infinite number that it is possible to construct. In particular the first given of the three is equilateral.

Taking the base square $ABCD$ as in fig. 126 we have to find compartment boundaries for the contacts 1 to 4, 2 to 3 that will

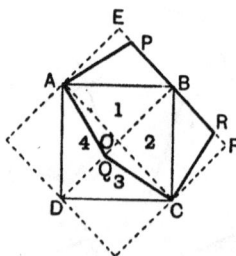

Fig. 126.

lie wholly inside the compartment square $AOBE$ and the three others similarly situated in regard to the other sides BC, CD, DA. In order that the resulting pattern may be a convex pentagon we choose any point P in EB and take a compartment boundary

APB for the compartment 1. This necessitates the choice of a similarly situated point *Q* on the line *OD* and the compartment boundary *AQD* for the compartment 4.

Thus the projection *APB* on one pattern will fit into the indent *AQD* in another.

We now select for the compartment 3, the boundary *DQC* (an indent). This selection necessitates the projecting boundary *BRC*, *BRC* being similar to *DQC* and the inverse of *APB*, for the compartment 2.

The result is the convex repeating pentagonal pattern *APRCQ*.

It will be noticed that *PAQ* and *RCQ* are necessarily both right angles.

We can thus obtain an infinite number of convex pentagons as repeating patterns, by simply varying the situation of *P* upon the line *EB*.

If *P* coincides with *B* the pentagon degenerates to the square *ABCD*.

If *P* coincides with *E* the pentagon degenerates to the rectangle *AEFC*.

In order to make the pentagon equilateral we first notice that by construction it has *four* sides equal, viz. *AP, AQ, QC, CR*. The outstanding side is *PR* in which *PB* is, by construction, equal to *BR*. We have therefore to make *AP* equal to twice *PB* to make all *five* sides equal.

For this to be so we find that the angle *PAB* must be (to the nearest minute) 20° 42′ and thence we find that the angles *APR, PRC* are each 114° 18′ and the angle *AQC* 131° 24′.

Laying off the angle *PAB* with mathematical instruments the pentagon is constructed.

This however is not necessary because we can readily make the construction with ruler and compasses only.

On a larger scale but with corresponding letters at various points, let *RBP* in fig. 127 be a side of the required pentagon, *B* its middle point, *BQ* a perpendicular and *BA* drawn so as to make an angle of 45° with *BP*.

With centre *P* and radius *PR* strike a circle to cut *BA* at *A*.

Then *PA* is a second side of the pentagon.

Similarly find the corresponding point *C* by the same construction to the left of *B*; *CR* is a third side of the pentagon.

With centre A and radius PR strike a circle cutting BQ in Q.

The remaining two sides are AQ, QC and the construction is complete.

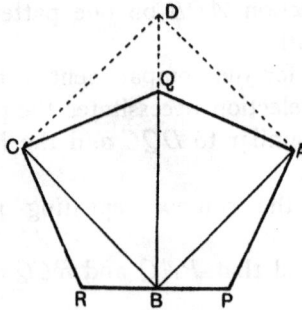

Fig. 127.

If the depicted pentagonal tessellation be studied it will be found to consist of a series of oblong hexagons in direction ⟋ cut by a second similar series in direction ⟍. This observation supplies the simplest mode of construction of the tessellation.

The area of this hay-stack shaped pentagon is equal to the square upon AB. If we produce BQ to D so that $QD = RB$ (half the length of the side of the pentagon) the figure $ABCD$ is the square upon AB.

This is very evident directly it is noticed that the triangles CRB, APB are equal and similar to the triangles CQD, AQD respectively.

Of 4-sided figures, besides the square we may conveniently take as bases the various forms assumed by the rhombus and rectangle. The reader should have no difficulty in dealing with these upon the principles set forth above. In the case of the rectangle he must observe that sides of unequal lengths must be associated with *different* colours when designing repeating patterns.

We may adopt the equilateral pentagon as a base upon which to construct a system of repeating patterns.

THE PENTAGON BASE

Some indeed there have been, of a more heroical strain, who striving to gaincope these ambages by venturing on a new discovery, have made their voyage in half the time.

JOH. ROBOTHAM to the Reader in Comenius's *Janua Ling.* Ed. 1659.

To take out other works in a new sampler.

MIDDLETON.

When the pentagon is assembled (see fig. 128) the contact system, as the reader will see at once on trial, is 1 to 1, 2 to 3, 4 to 5.

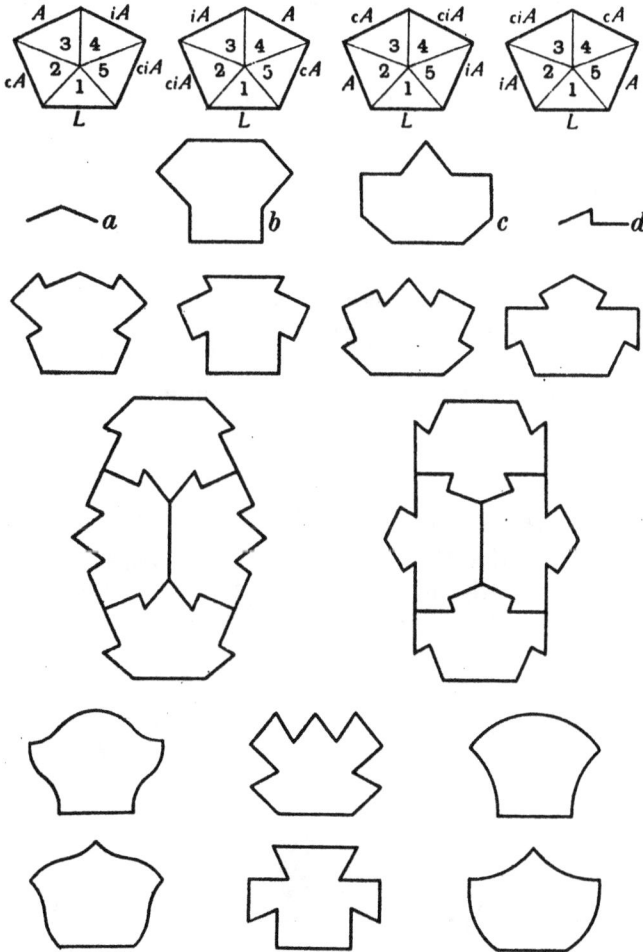

Fig. 128.

7—5

The side being 20 millimetres, the altitude is 26·5 mm. and the point upon the central axis to which the angular points are joined is 11 mm. from the base.

The five compartments thus formed have no angle less than 45°.

The labelling of the sides shews, for any appropriate boundary A, four distinct symmetrical patterns.

The sides labelled A, iA' in the first figure are inclined to the horizontal line at the same angle that the sides labelled cA, ciA are inclined to the vertical.

The angle in question is approximately 24° 18′.

This fact leads us to take for A in the first place the boundary as in fig. 128 a; easily constructed because the tangent of 24° 18′ is ·45. The base angles of the boundary are thus each 24° 18′.

The four resulting patterns degenerate to two, as in fig. 128 b and c, because $A = iA$ in this special case.

Next take for A the form fig. 128 d and we obtain the full number of patterns as shewn in the third row.

> So to their work they sit, and each doth chuse
> What story she will for her tapet take.
>
> SPENSER, *Muiopotmos.*

I went alone to take one of all the other fragrant flowers that diapred this valley.

GREENE'S *Quip for an Upstart Courtier*, B. 2.

THE REGULAR HEXAGON BASE

82. Numbering the compartments, as before, as in fig. 129 a, there are apparently only three possible arrangements. One is shewn in fig. 129 b in which compartments 1, 2, 4, 5 are adjacent

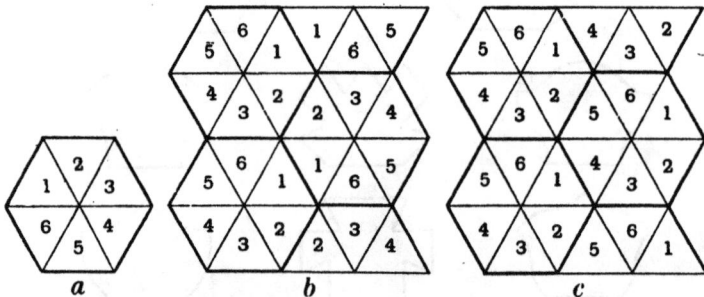

Fig. 129.

to 1, 2, 4, 5 respectively and 3 adjacent to 6 The piece has two aspects.

A second is shewn in fig. 129 c according to the contact system 1 to 4, 2 to 5, 3 to 6. The piece has one aspect.

In these systems numbers which are not linked with different numbers may be made identical to any extent ; also two numbers linked together may be made identical and numbered at pleasure without interfering with the assemblage. The particular cases are too numerous to be separately examined in this book.

The patterns and assemblages which are shewn in fig. 130 belong to the first of these contact systems. The assemblages exhibit each two aspects.

Nicophanes gave his mind wholly to antique pictures, partly to exemplify and take out their patterns.

HOLLAND'S *Pliny*.

The patterns and assemblages which are shewn in fig. 131 belong to the second of these contact systems. The assemblages have each one aspect.

Take me this work out.

Othello, III. iv.

83. When the contact system is 1 to 1, 2 to 2, 3 to 6, 4 to 4, 5 to 5 the symmetry may be either about an axis perpendicular or parallel to a side. In the former case, we have for one boundary form of the first kind the set shewn in fig. 132 *a, b, c, d*.

Taking for A the boundary as in fig. 132 *e* and for B as in fig. 132 *f* we find the patterns given in fig. 132 *g, h, k, l*.

If the symmetry be about a diagonal we have the cases of fig. 132 *m, n, o, p*.

Employing two boundary forms of the first kind B, C we have fig. 132 *q* and seven others by replacing B and C by B, iC; iB, C; iB, iC; C, B; C, iB; iC, B; iC, iB respectively.

We also have fig. 132 *r* and three others by replacing B and C by B, iC; iB, C; iB, iC respectively.

84. When the contact system is 1 to 4, 2 to 5, 3 to 6 we have symmetrical arrangements when the axis of symmetry is perpendicular to a side and when parallel to a side. Thus we have fig. 133 *a* and three others obtained by replacing B by cB, iB, ciB. We also have fig. 133 *b* and *c* for the diagonal symmetry.

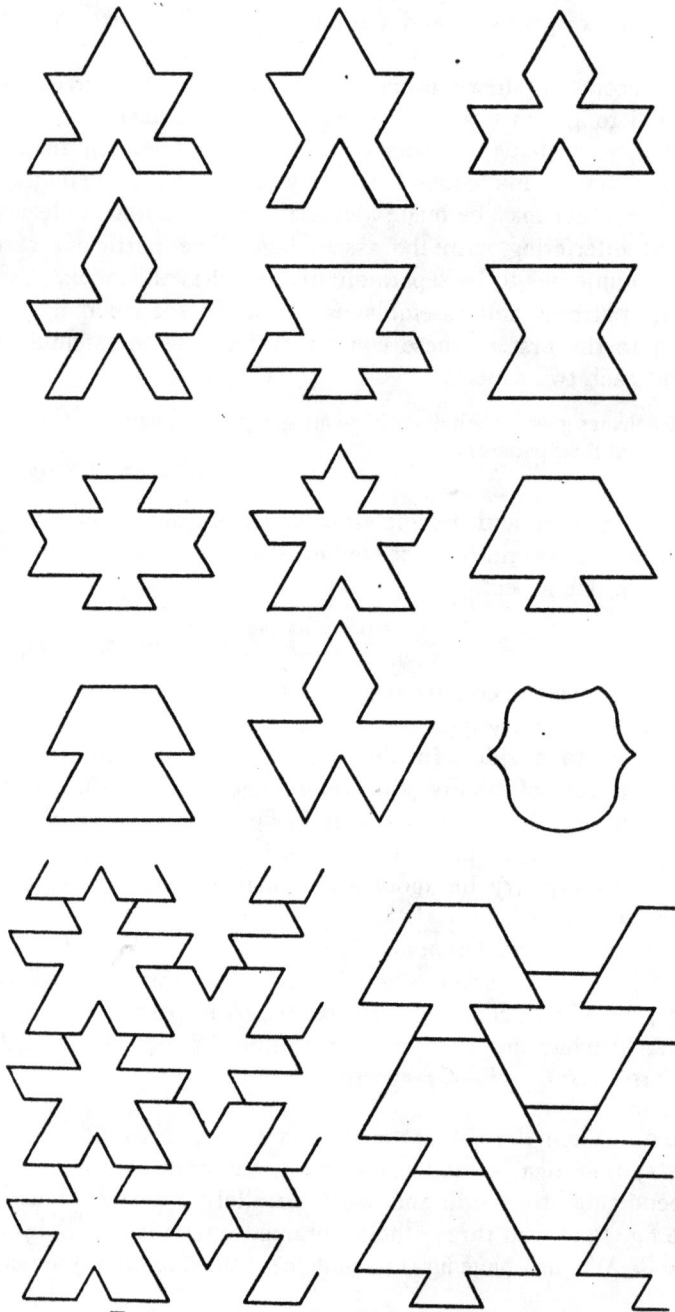

Two aspects   Fig. 130.   Two aspects

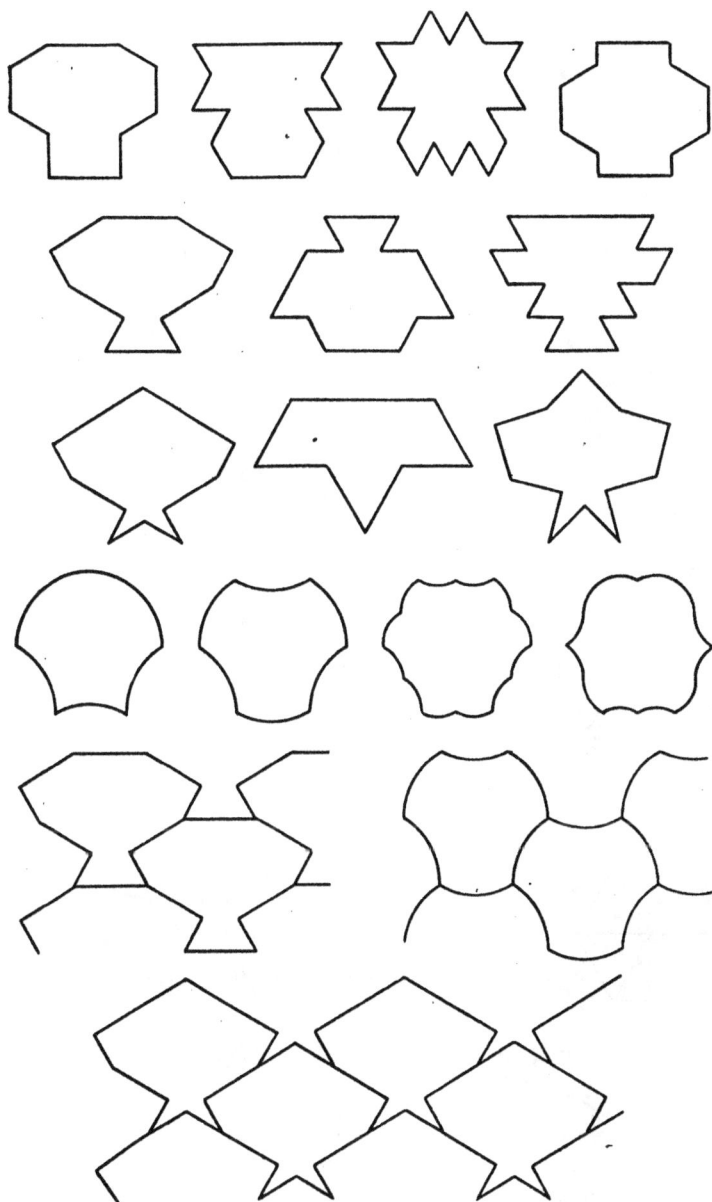

One aspect

Fig. 131.

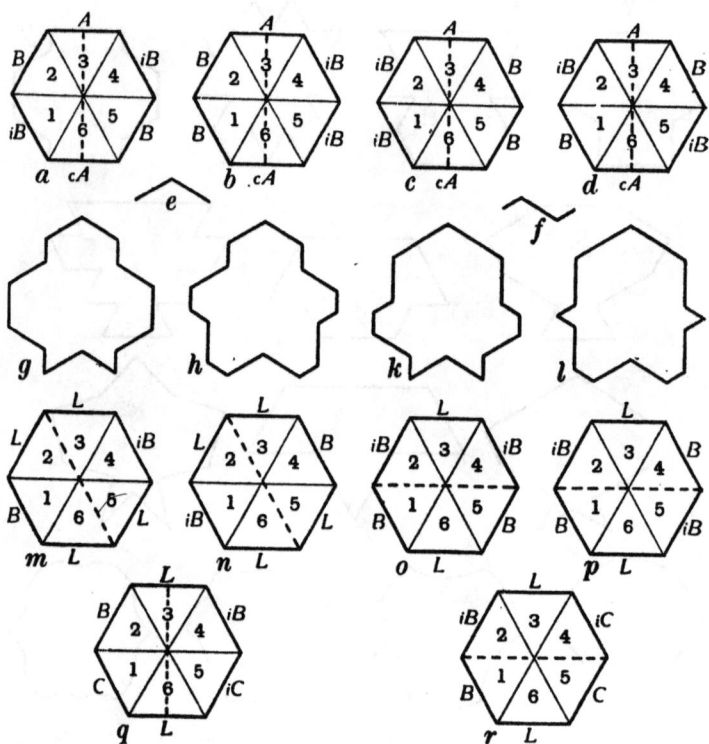

Fig. 132.

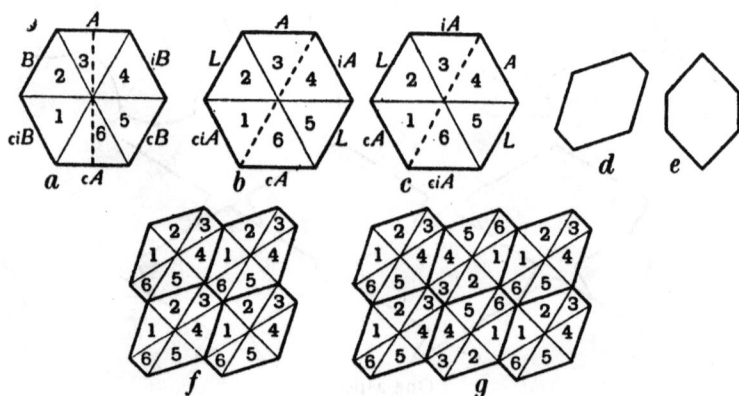

Fig. 133.

L is suitable because it is the simplest form of the second kind which is both its own complement and its own inverse.

Other hexagons may be taken as bases; such for instance as arise from the square base and are shewn in fig. 133 *d* and *e* giving assemblages as in fig. 133 *f* (1 to 4, 2 to 5, 3 to 6) and as in fig. 133 *g* (1 to 1, 2 to 5, 3 to 3, 4 to 4, 6 to 6). As with the regular hexagon two contact systems are available.

The third contact system 1 to 2, 3 to 4, 5 to 6 is established by the diagram shewn in fig. 134 *a*, which exhibits three aspects.

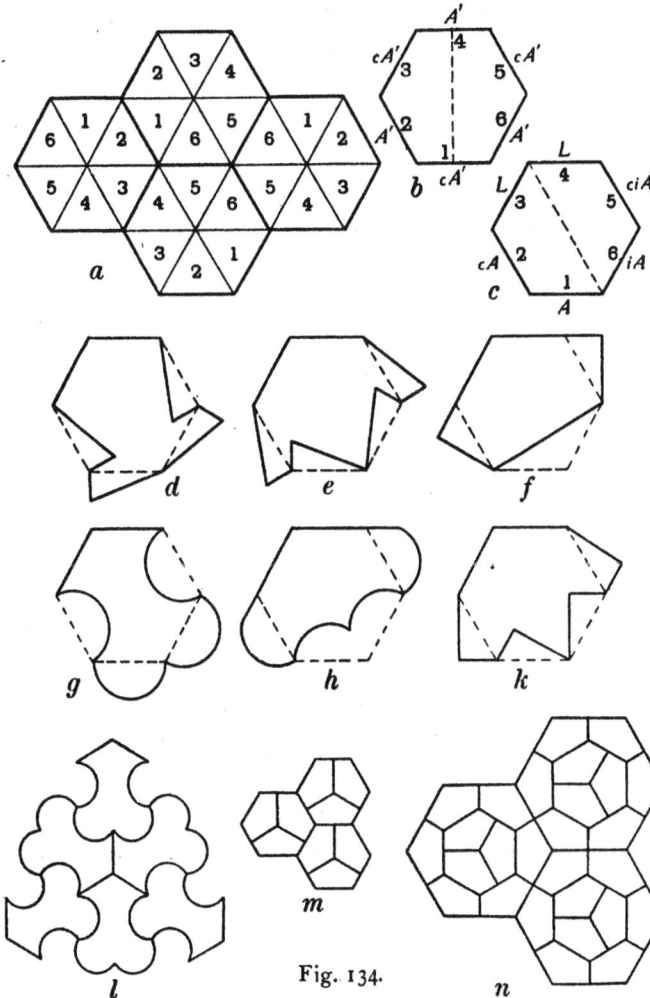

Fig. 134.

Symmetry of pattern can be obtained either about a line bisecting opposite sides or about a line bisecting opposite angles as in fig. 134 *b* and *c* respectively.

In the first diagram *A'* is a self-inverse boundary.

In the second diagram *A* is any boundary.

L as usual is the unaltered straight line.

The first of these does not require attention because it is derivable from the contact system 1 to 4, 2 to 5, 3 to 6 which has been already considered.

But giving *A* various forms, as in fig. 134 *d* to *k*, in the second we find assemblages of which those shewn in fig. 134 *l, m, n* are examples.

Exceptionally the pentagon pattern, which appears, can be assembled so as to exhibit either three (fig. 134 *m*) or six aspects (fig. 134 *n*).

It will be noticed that the pentagon is a repeating pattern on the principle given early in this Part—that if a repeating pattern can be dissected into a number of parts of the same size and shape, the shape involved is also a repeating pattern. Here the regular hexagon is divisible into either 3 or 9 similar and equal pentagons, the pentagon being the one before us.

85. The equilateral triangle, the square, and the regular hexagon are satisfactory bases for the construction of repeating patterns because they are themselves repeating patterns. In fact any repeating pattern may be taken as a basis for the evolution of other repeating patterns. This general principle will occasionally and exceptionally be found of interest and importance.

In order to throw some light upon this development we will consider a very well known repeating pattern for linoleum, pavements, etc., viz. the combination of the regular octagon and the square (fig. 135 *a*).

How can we obtain this form upon the principles that have been set forth?

When, as in this case, we are given a repeating pattern and we wish to determine the base and contact system to which it appertains, we have a problem which sometimes requires cleverness and ingenuity for its solution.

Take a square base as in fig. 135 *b* and the contact system 1 to 3, 2 to 4 with the boundaries as in fig. 135 *c*.

The result is the repeating pattern of fig. 135 *d*.

Numbering its compartments, take it as a new base as in fig. 135 *e* with the contact system 1 to 4, 2 to 5, 3 to 6 and with the boundaries as in fig. 135 *f*.

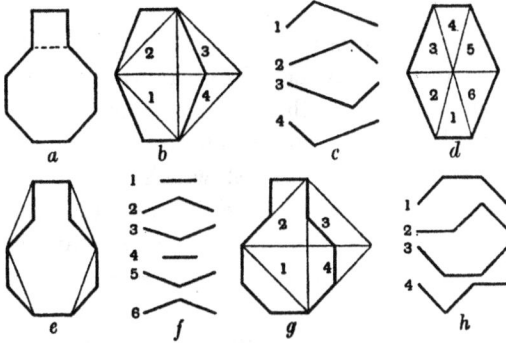

Fig. 135.

The result is the repeating pattern (fig. 135 *a*) of which we are in search. It has emerged by combining a square base and the contact system 1 to 3, 2 to 4 with an hexagonal base having the contact system 1 to 4, 2 to 5, 3 to 6.

We now find that the pattern may be made to emerge at once from the square base as in fig. 135 *g* with the contact system 1 to 3, 2 to 4 by taking the boundaries as in fig. 135 *h*.

In colouring assemblages of repeating patterns it is a useful rule to employ at least as many colours as the pattern has aspects.

Usually each aspect would be associated with a separate colour, but this need not be the invariable rule.

THE CONSTRUCTION OF THE PASTIMES

> Thou art a three-pil'd piece, I'll warrant thee.
> *Meas. for Meas.* I. ii.

86. In Part I the triangles may be made at home out of stout cardboard and have a side two-and-a-half or three inches in length. The compartments should be coloured with good water-colours or oils. Four suitable colours are black, white, red and blue, as they are readily distinguishable at night. A more permanent set should be made of good dense wood. The writer's set has a three-inch side and the thickness is one-fifth of an inch.

Oil colours are used and for the five-colour set the colours are black, white, red, olive green and dark orange. If made at home, in cardboard, pains should be taken to make the shape as accurate as possible. The pleasure of handling the pieces is much increased when they fit accurately. The ideal pieces would be as heavy as ordinary dominoes. The heavier they are the better.

The square pieces may be of one-and-a-half or two inches side and one-sixth or one-fifth of an inch thick.

It is desirable to have boards upon which to assemble the pieces, boards with a rim against which the boundary pieces may be placed. For the equilateral triangular pieces, assuming a piece of three inches side, the board should be a regular hexagon of slightly over six inches side so that the pieces will just fit in comfortably.

For the 20-set 5-colour triangular pieces the board should be the figure-of-eight shape as depicted. For the square pieces the only board required is the 6 × 4 board and for the right-angled triangles the shape of the hexagon used.

The irregularly shaped pieces of Part II and the designing of Part III will be much facilitated by the use of squared millimetre paper. This is preferable to squared paper in subdivisions of the inch because, at any rate in this country, it appears to be of more reliable accuracy.

The pieces should be set up with some simple boundary and then drawn on the millimetre paper. This can then be pasted upon 3-ply or other suitable wood and handed over to some one with a fret-saw to cut out. The pattern-maker should be called in if sets of great accuracy are required, and they should be as heavy as possible.

BIBLIOGRAPHY

BACHET DE MESIRIAC. Problemes plaisans et delectables qui se font par les nombres. 1612. A. Labosne. Paris, 1884.

LEUVECHON, JEAN. Récréation mathématique, etc. 1624.

LEAKE. Mathematical Recreations. London, 1653.

OUGHTRED, WILLIAM. Mathematical Recreations. London, 1674.

OZANAM, JACQUES. Récréations mathématiques, 1694. English translation by Charles Hutton. London, 1814.

BERCKENHAMP, J. A. Les amusements math. Paris, 1749.

HOOPER, W. Rational Recreations. London, 1774.

ALLIZEAU, M. A. Les métamorphoses ou amusements géométriques. Paris, 1818.

JACKSON. Rational Amusements for winter evenings. London, 1824.

HUGONLIN. Première collection de récréations mathématiques. Paris, 1828.

ROBINSON, N. H. Mathematical Recreations. Albany, 1851.

LECOT, V. Récréations math. Paris, 1853.

LA GARRIGNE, F. Curiosités math. Clichy, 1874.

CANTOR, M. Zahlentheoretische Spielerei. Zeitschr. für Math. 1875.

TISSANDIER, GASTON. Les Récréations math. ou l'enseignement par les jeux. Paris, 1881.

LUCAS, ED. Récréations math. Paris, 1882—1894.

HÉRAUD, A. Jeux et récréations scientifiques. Paris, 1884-1903.

RIVELLY, ALF. I giuochi matematici. Napoli, 1887.

LUCAS, ED. Jeux scientifiques pour servir à l'histoire, à l'enseignement et à la pratique du calcul et du dessin. Paris, 1889.

LEVY, LUCIEN. Sur les pavages à l'aide de polygones réguliers. Bull. Soc. philomath. 1890.

LUCAS, ED. Théories des nombres. Paris, 1891.

BALL, W. W. ROUSE. Math. Recreations and Problems. 5th ed., 1911.

SUNDARA, ROW. T. Geometrical Exercises in Paper Folding. Madras, 1893.

VINOT, J. Récréations math. Paris, 1893.

LUCAS, ED. L'arithmétique amusante. Paris, 1895.

SCHUBERT, H. Math. Mussestunden. Leipzig, 1898.

FOURREY, E. Récréations math. Paris, 1899.

AHRENS, W. Math. Unterhaltungen und Spiele. Leipzig, 1901.

IGNATIEV, E. J. Math. Spiele, Rätsel und Erholungen. Petersburg, 1903.

TEYSSONNEAU, E. Cent récréations math.; curiosités scientifiques. Paris, 1904.

FOURREY, E. Curiosités géométriques. Paris, 1907.

DUDENEY, H. E. The World's Best Puzzles. Strand Magazine. 1908.

DUDENEY, H. E. The Canterbury Puzzles. 1908.

AHRENS, W. Mathematische Unterhaltungen und Spiele. Leipzig. Aufl. I,
    1910; II, 1918.

ERNST, E. Math. Unterhaltungen. Ravensburg, 1911–12.

GHERSI, ITALO. Matematica dilettevole e curiosa. Milano, 1913.

GENAU, A. Math. Überraschungen. Arnsberg, 1913.

DUDENEY, H. E. Amusements in Mathematics. London and New York, 1917.

AHRENS, W. Altes und Neues aus der Unterhaltungsmathematik. Berlin,
    1918.

> And now he has pour'd out his ydle mind
> In dainty delices and lavish joys.
>
> SPENSER, *F. Q.* II. v. 28.

PRINTED IN ENGLAND BY J. B. PEACE, M.A.
AT THE CAMBRIDGE UNIVERSITY PRESS

Further Reading and Contacts

Tarquin Reprints

Tarquin Reprints is a new series of books that endeavours to bring back to life important books in mathematics. An expert editorial team have identified the first works to be so treated – but other suggestions are very welcome. Contact Andrew Griffin at qed@talk21.com with your ideas. Further information on new titles at the web address below.

New Mathematics Pastimes CD

This book has a CD-ROM with a series of additional resources and a colour version of many of the diagrams, prepared by John Sharp. To obtain this, go to www.mathsite.co.uk and search for MacMahon.

www.mathsite.co.uk

For mathematics resources from Tarquin, QED, Dover, Key Curriculum Press, NCTM, Efofex and many more – mathsite is your one stop shop to buy in £, $, € for despatch anywhere across the world.

Recreational Mathematics
Putting the fun back into maths!

Tarquin is delighted to announce its new magazine for adult readers – launching early in 2005. Coverage is from leading authors under an editorial board led by John Sharp – and includes:
- Maths and art
- Paper engineering
- Number
- Puzzles
- History of maths
- And much more!

Register for a free copy on www.mathsite.co.uk.